the green blue book

the green

blue

book

the simple water-savings guide to everything in your life

thomas m. kostigen

RODALE

© 2010 by Thomas M. Kostigen

Rodale books may be purchased for business or promotional use or for special sales. For information, please write to:

Special Markets Department, Rodale Inc., 733 Third Avenue, New York, NY 10017

Printed in the United States of America

Rodale Inc. makes every effort to use acid-free ⊗, recycled paper ♻.

The interior of this book was printed on Cascade Enviro 100% PCW, FSC certified paper; the cover is 100# Mohawk VIA Smooth PCW Cool White FSC certified; and the book will be printed using soy inks.

The amount of water used in the production and manufacturing of The Green Blue Book has been replenished by the Bonneville Environmental Foundation's Water Restoration Certificate Program, which restores water to critically de-watered ecosystems.

For more information on this innovative process and how to mitigate your water use, visit www.BEFWater.org.

Recycled
Supporting responsible use of forest resources
FSC
www.fsc.org   Cert no. SW-COC-002550
© 1996 Forest Stewardship Council

100%

Library of Congress Cataloging-in-Publication Data
Kostigen, Thomas.
   The green blue book : the simple water-savings guide to everything in your life / Thomas M. Kostigen.
      p.   cm.
   Includes bibliographical references and index.
   ISBN-13 978-1-60529-471-1 paperback
   ISBN-10 1-60529-471-3 paperback
   1. Water conservation.   I. Title.
   TD388.K67 2010
   333.91′16—dc22   2009052250

Distributed to the trade by Macmillan
2 4 6 8 10 9 7 5 3 1 paperback

RODALE
LIVE YOUR WHOLE LIFE™

We inspire and enable people to improve their lives and the world around them
For more of our products visit **rodalestore.com** or call 800-848-4735

for the girl in the purple dress

contents

introduction   ix

section 1: the water you see

1: home
   home is where the water is   3

2: outdoors
   trimming, timing, and topping off   13

3: work
   make it a water recession   21

4: sports
   keeping a low water score   27

5: travel
   away doesn't mean a waste   33

section 2: the water you can't see

6: foods and beverages
   how much does it take to produce your produce?   43

7: clothing
   are you wearing water?   67

8: furnishings
   the H_2O of household items   73

9: health and beauty
   bail out your bathroom   79

10: school and office products
    the (water) supply closet   85

11: luxury
   bling bilge   91

12: pets
   the irony of fire hydrants   97

13: building materials and appliances
   it takes a lake to lay a foundation   103

section 3: the sum of your water existence

14: the water footprint calculator
   liquid math   113

conclusion   125

for more information and resources   131

references   133

quick guide   189

acknowledgments   197

index   199

introduction

The most important thing that you can do to save the planet right here, right now, is to save water. In fact, former vice president Al Gore now says during his famous *Inconvenient Truth* slide-show presentations that "water is the cutting edge of environmentalism." Why? Because the world, which has had almost to the drop the same amount of water on it since dinosaurs roamed, is quickly running out of its available freshwater—what we use for drinking, bathing, and irrigating. And that means life is about to change dramatically for all of us.

In the United States, we're already seeing the water crisis encroach: Wildfires, droughts, and rationing are spreading throughout the country. We're growing fewer crops and raising fewer livestock. With less water comes less food, and the chain of effects gets real serious real fast. And that's here in relatively prosperous America. Current trends of overconsumption, pollution, and climate change are leading to a situation where two-thirds of the global population will face severe water shortages by 2025.

Water is a complicated case of supply and demand. Population growth has a lot to do with demand and the water crisis—having more people creates more demand. At the same time, global warming, pollution, and mismanagement are limiting our supply.

Freshwater makes up less than 1 percent of the total volume of water on Earth. Better decisions have to be made about its use.

The Green Blue Book: The Simple Water-Savings Guide to

Everything in Your Life will show you how to use water wisely—simply, without sacrifice, through better decision making. The hope is that this will have a trickle-up effect and merge with a larger movement of water conservation and management that includes businesses and governments.

I have traveled the world reporting and writing about environmental issues. And sure, climate change is a major problem, pollution is a problem, and waste is a problem. But water, well, we can't live without it—not for more than about 3 days, anyway.

What I find so intriguing about the water problem is that it can be solved with attention. International treaties can incorporate water into trade equations, new technologies can increase supply, and we can do our part. We can make change happen now, each and every one of us, with the choices we make in our lives.

They add up.

For example, if we more accurately measure the amount of water we use to brew coffee and each of us saves just 1 cup from getting tossed down the drain, in a year we'll save enough to provide 2 gallons to every one of the 1.1 billion people who don't have access to freshwater at all.

Simple action. Big consequence. Changing our habits can be interesting and empowering.

Of course, making less coffee will make a difference—if we can get everyone to do it. However, choosing the *type* of coffee to buy will make the most difference. The water it takes to grow the coffee outweighs the water it takes to make a cup.

The concept of how much water it takes to make or grow something is called its "virtual water" content. The idea is that once you know the difference between items in terms of how much water they require, you can make choices to save. It takes about 123 gallons of water to grow a pound of oats, for example. So the virtual water count of this pound of oats is 123 gallons. Now, extend this to a slice of beef. Since cattle have to feed on oats or some other grain and obviously need to drink water them-selves, their virtual water count is higher than that of what they chew. It also takes water to process their meat. The virtual water footprint of a pound of beef, therefore, is 1,500 gallons. When you understand that there's a virtual water count to producing every-thing—a beer (20 gallons), a glass of wine (30 gallons), a cup of coffee (37 gallons), a cup of tea (5 gallons), a car (39,000 gallons), a bicycle (480 gallons)—you can make water-smart choices. This is what I hope *The Green Blue Book* will be: a great resource for understanding the virtual water content of the things you con-sume, as well as ways to save.

In the United States, the average person uses more than 656,000 gallons of water per year. The world average is about half that. Clearly, we can stand to be put on a water diet.

Seventy percent or more of the residential water use in some parts of America is for our lawns. We waste up to 5 gallons of water when we flush our tissues down the toilet instead of tossing them in the trash can. Here, waste is the key. It isn't about doing without water, it's about not wasting it—or polluting it. Some 90 cities still dump raw sewage into the Great Lakes, our largest source of freshwater and fully 20 percent of the world's surface freshwater supply. Individual actions along with conservation and clean-water policies need to change.

Life on Earth depends on the global water cycle, which is in constant motion: Water evaporates from the oceans to form clouds, which cruise across the sky to deposit rain or (sometimes) snow on the land. Much of this freshwater eventually flows back to the ocean via streams, rivers, and storm drains, while a smaller percentage gets tied up in vegetation, buried in a glacier, or trapped in an underground aquifer until it is released back into the moving cycle via transpiration, snow melting, or your garden hose running. This water cycle has not only been functioning effectively for eons, but it also has no identifiable beginning or end. Simply stated, water molecules—made up of two hydrogen atoms and one oxygen atom (H_2O)—don't disappear. They may change state, move from place to place, or get incorporated into more complex molecules, but the water balance on this planet has stayed relatively constant for millions of years.

Enter *Humana interruptus*. When water is used inefficiently or wastefully, the result can be less water in regions where clean freshwater is already scarce.

By minding our water consumption, we can save those who are suffering, or are about to suffer, from a more arid planet. And I'm not strictly talking about Ethiopians and the like. Thirty-six US states will experience droughts over the next 5 years. California is rationing its water supply. Arizona has been forced to import its water. A couple of years ago, Georgia almost ran out of water altogether. Texas, the second biggest agricultural state behind California, is now the driest region in the country. The Great Lakes, which recently reached historic lows, are even fending off water poachers. And that's just here in the United States.

The Himalayan glaciers, which supply freshwater to half the world's population, are expected to melt away within the next 50 years. Water worldwide is an increasingly big concern.

The Green Blue Book can help to shore up supplies. Through a comprehensive listing of products, actions, and choices, this book shows how we can all save water and play a role in controlling the crisis afoot. It also includes a water calculator so you can figure the total amount of water you use. This total amount is called your water footprint. It takes into account home water use and the amount of water used to produce the food you eat and the products you buy. It also includes other factors such as the water saved by recycling and composting. Along with water savings, it shows you how to save money in ways you may not have thought of before: trimming that water bill, for instance.

The Green Blue Book isn't meant to be preachy; it's practical. Hundreds of products' virtual water contents are featured for you to see. Dozens of actions showcase easy water-saving measures. And water-saving tools (such as rain sensors, low-flow plumbing attachments, and home water monitors) are listed to empower you to save.

Pick it up, flip to a page, and get the scoop on how much water it takes to make a piece of paper, a slice of bread, or a pair of shoes. Flip again and find out how to save thousands of gallons of water by using a dishwasher rather than washing by hand. Flip yet again and learn how to calculate all the water you use in a year, as well as to figure out ways to lessen your usage.

To be sure, the solutions you'll find in this book are meant primarily to create water savings. However, it should be understood that a general mindfulness about the environment is woven throughout. Where appropriate, I explicitly note when other environmental savings are being sacrificed for water savings. That said, practices such as organic farming and healthy eating more often than not also save water. That's yet another rationale for being health conscious and environmentally friendly: It all ties together.

I wrote *The Green Blue Book* because there wasn't a book like it that was fun to read and showcased comprehensive, actionable steps to help fight the water crisis we are facing.

Now, hopefully, there is.

section
1

the water you see

better ways to manage use

home

home is where the water is

I used to daydream in the shower, letting my mind fog over like the bathroom mirror fogged with steam. Then I became aware of how much water I was wasting with every extra minute I stood there. Now it's as if I shower at the Bates Motel: I'm in and out quickly.

Many of us mindlessly waste water, either because we are just fogging out or because we really don't know any better. A home-based education on water is what we need. And here's why: Our homes are where we use the most water in our lives. The average household in America uses about 400 gallons of water per day. That can easily be cut to less than 100 gallons by doing a few simple things.

But first I'd like to explain what the average home looks like. Sixty-thousand-square-foot mansions like Bill Gates's home (which is rumored to use about 5 million gallons of water a year, more than 30 times the average home's use, by the way) aren't what I'm describing. Rather, the average size of a new home in the United

States is more than 2,500 square feet—780 square feet larger than the typical home built 3 decades ago. Picture a ranch-style home, which is uniquely American. It has 3 bedrooms, 2½ bathrooms, a laundry room, and a 300-square-foot kitchen. The direct sources of water inside this house include fixtures, pipes, tubing, and toilets. From these taps all sorts of routine activities spawn. We brush our teeth, shave, wash dishes, shower, take baths, do laundry, bathe babies. But how much thought do we put into the means of our doings compared to how much we think about the action itself? We are careful not to cut ourselves shaving, but we usually aren't so careful about not leaving the tap running.

Our homes provide us with sustenance, yet we take much of this for granted. Water, especially, is expected. Until the mid-1800s, water wasn't commonly available via indoor plumbing. People pumped it from underground wells or lived by rivers. It wasn't until the 1930s that the federal government started declaring a home substandard if it didn't have indoor plumbing.

The Romans may have built an amazing municipal water and sewer system as far back as 800 BC, but it took centuries for people to get used to flushing, washing, and drinking in the privacy of their own homes. This was mostly due to cost. The cost for a private home to install indoor plumbing in the late 19th century was more than $500—an enormous sum at that time. Back then the average home was less than half the size it is today.

Size, you might say, doesn't matter when it comes to water use. After all, an additional bedroom doesn't necessarily equate to more water consumption. But more bathrooms do. So do bigger kitchens. We've learned to pipe in more to feed our water frenzy in the name of convenience. Not that there is anything wrong with convenience. It's a marvel to turn on a faucet and have your choice of hot or cold water. Imagine just how awestruck a caveman would be by this magic.

Consciousness shouldn't have to take a backseat to luxury, however. We can lessen the profligate drain on our water supplies without sacrificing much, if anything. The average person needs about 13 gallons of water a day to drink, wash, and eat. We in America use almost 10 times that. In fact, the global population may have tripled in the 20th century, but water consumption went up sevenfold.

Using less doesn't mean we have to go backward in time, or do without anything. It means using water more wisely. And here's how.

IN THE BATHROOM . . .

Turn off the tap when you brush your teeth. Do as your dentist recommends and brush your teeth three times a day, and, tap running, you'll likely use about 5 gallons of water. Turn off the tap, and you could use as little as a few tablespoons. If every American did this, the savings after just 1 day would be as much as all the residents of an average-size state use in 2½ days.

Install a low-flow or dual flush toilet. These types use less water per flush, and dual flush models give you the choice of flushing with more water or less depending on the, um, "action." If you can't install a new toilet, put a half-gallon jug of water in your tank—make sure it doesn't float—to reduce its volume. Toilets account for about 27 percent of the water we consume indoors. As a nation, we flush almost 11 billion gallons per day. Place a half-gallon jug in your tank, and your household of four could save about 4,400 gallons per year. Low-flows and dual flushes cut use in half—up to 3 gallons per flush—compared to older toilets.

Take a shorter shower. The average person showers for
8 minutes a day. Every minute uses about 2½ gallons of
water. So that's 20 gallons of water for the average shower.
If everyone turned off the water 1 minute sooner, the savings
across the country would total nearly 12 billion gallons.

Screw on a low-flow showerhead. These and low-flow faucet
aerators (see page 9) can reduce your water consumption by
50 percent. That means your teenager can still spend twice
as much time in the shower—but only use half the amount of
water.

Skip baths. Filling the tub can use three times as much water as
taking an average-length shower. This savings alone amounts
to 12,000 gallons a year. That's about as much water as it
takes to fill an average above-ground swimming pool.

Fill the sink instead of letting the water run when you shave.
You'll save 90 gallons of water a month. We could fill a
football stadium 1,370 times with the water saved if every US
adult male did this.

IN THE KITCHEN . . .

**Measure your coffee-making water more accurately, and the
1 extra cup you leave behind at the bottom of the coffeepot
won't go to waste.** The biggest single use of drinking water is
for making coffee. The 48 percent of US adults, or 106 million
people, who drink it every day can make sure it's good to the
last drop. Savings: 2.4 billion gallons per year.

Don't always believe the box when you cook. It only takes
1½ quarts of water, for instance, to cook a pound of pasta,

whereas most instructions say it takes between 4 and 6 quarts. Considering that we cook a billion pounds of pasta per year in the United States, the water savings could equal a billion gallons as well.

You don't have to drink that much water to be hydrated. It's a myth that you need eight 8-ounce glasses of water every day to stay healthy. Four 8-ounce glasses will do for the average healthy human being. And even that is generous: We get most of the water we need to survive from the food we eat. Cut the myth in half, and we'd save more than 300 million liters of water a day, which is almost as much as all the bottled water sold every day throughout the world.

Only freeze ice cubes you'll use within a month, especially if you have a frost-free freezer. The water "in" ice cubes evaporates after several weeks because of sublimation, when H_2O goes from its solid state to a gas. Your freezer fan and opening and closing the freezer door speed up this process. It takes about 1½ cups of water to fill a tray. Evaporate a tray from every household in America, and we lose some 10 million gallons into thin air.

Compost instead of running water to clear your disposal. Five gallons of water per minute are being wasted as your food goes down the drain. Compost by turning your food waste into fertilizer and at the same time save almost 2,000 gallons of water per year.

Soak your vegetables instead of rinsing them with the tap on. You'll need only about a cup of water for a pound of vegetables versus the 80 cups you'd use by letting the tap run.

Have designated drinking glasses for family members that they can use throughout the day—or across multiple

days—instead of having them take a new glass each time they get thirsty. This will cut down on the washing (and the water) required for no good reason except a "fresh glass."

Scrape excess food into the trash can or compost bin instead of rinsing dishes before placing them in the dishwasher. Why rinse-wash-rinse when you can just wash-rinse? Sticky food lovers: Wet a sponge and use that instead of letting the tap run.

Use your dishwasher, and you'll use half as much water as you would washing by hand. The average dishwasher uses about 11 gallons of water per load. At four loads a week, you'll save 2,300 gallons a year.

IN THE LAUNDRY ROOM . . .

If you wash clothes by hand, don't. It's the 21st century. Washing and rinsing just one garment by hand can use as much water as a whole load in an efficient washing machine: 20 gallons.

Use a front-loading Energy Star washing machine and cut down on the water used for your wash cycles. A front-loader uses 13 gallons less, on average, than the typical top-loading washer. Considering that there are 80 million washing machines in the United States, more than 400 billion gallons of water per year could be saved. If you are stuck with a top-loader for now, reduce your wash times. Depending on the machine, you can save about 5 gallons per minute.

Cut down on the number of loads you do. Wash full loads and you'll achieve maximum efficiency with your washing machine. A typical household uses 15,000 gallons of water to wash clothes each year, which is about 15 percent of the total indoor water use. Double your loads and save 7,500 gallons of water a year. If you were to bottle the savings, you could fill more than 28,000 liter-size containers, more than enough for everyone in a small town.

IN THE LIVING AREAS . . .

The number one plant killer is overwatering. Professional gardeners claim that most indoor plants are overwatered by 90 percent. Most plants only need to be covered in 1 inch of water per week. The savings from just one plant in every household could flood the country.

Check the house for leaks. A dripping faucet can waste up to 20 gallons of water per day. That's as much as 7,300 gallons a year. A leaky toilet can waste up to 100 gallons a day, or 36,500 gallons a year! To check the toilet, put a drop of food coloring in the tank at night. If by morning it's in the bowl, you've got a leak. To check the faucet: Look or listen.

Use low-flow faucet aerators. They're not expensive and are easy to attach. A family of four can save more than 8,000 gallons of water per year, or roughly a whole summer's worth of showers.

Choose a refrigerated air conditioner instead of an evaporative cooler. Evaporative units cool air by bringing it into close contact with water and can account for 30 percent of total annual water consumption if they're used constantly in the warmer months. (They can require up to 10 gallons or more an hour.) Air conditioner units use refrigerants to cool the air. These use more energy, but they save water.

Go tankless. Tankless water heaters (also called on-demand water heaters) instantaneously save water because they heat water directly without the use of a storage tank. Considering that it can take a minute or two of running the tap before hot water from a traditional water heater gets to you, an awful lot of waste—at least 5 gallons per awaiting-hot-water episode—can be avoided. Tankless water heaters are also up to 50 percent more energy efficient.

Water meters, just like gas and electric meters, sometimes lie. Check the reading on yours (it's typically located in an easy-to-spot place on the outside of your home), then fill a bucket with a few gallons to see if the reading is accurate. For obsessive types, smart water meter monitors have been developed that transmit signals to a little screen you can place inside your house (there are even little magnet ones for your refrigerator door). It's simple to determine individual use: Divide the total amount of water used by the number of people in the house. Other monitors come equipped with leak sensors and send out an alert. Yup, a blue alert!

Wrap your pipes. Installing piping insulation is a relatively easy fix for preventing hot water from cooling off on its way to the tap. When you get the water temperature you want sooner, you waste less. Insulation is inexpensive and can be found at most hardware stores. If you are in for the big fix, think about a direct plumbing plan. Long, winding pipes are inefficient. Brain teaser: Why is it that when you turn on the hot water, the volume of cold that comes out before the water warms up is one and a half times the volume of the pipes? You first have to push out all the cold water in the pipe before it flows hot. Then, once the hot water enters the pipe, you lose some of the initial heat to the cold pipe itself.

BOTTOM LINE

Check the time. Every minute a faucet runs, you use several gallons of water. If it's an older device, you're likely using three times that amount. No matter, get a low-flow device and save about half the water you'd normally use.

More accurately measure the amount of water you use. Whether it was for a plant, some pasta, or a pot of coffee, leftover water can't be saved for another day; it's wasted. Bad math is bad for the planet. Count the drops correctly.

Embrace technology. Sure, you save more when you use a dishwasher versus washing by hand, and the same holds true for your car; go to an automatic car wash and save 100 gallons compared to your own hose-and-bucket job.

2

outdoors

trimming, timing, and topping off

Outdoors is where we as residents tend to use huge amounts of water. In some parts of the country, mostly out in the arid West, 70 percent or more of residential water is used for lawn irrigation. Something is seriously wrong with this picture. Pink flamingos and fountains aside, decorative lawns that need lots of care and lots of water are scourges. It may be that suburbia is making the wells run dry. Indeed, homeowners use an average of 120 gallons of water each day for things outside. Think about that for a second: "things outside" . . . where rain should be able to do the job nicely . . . if we stick with the vegetation that grows naturally in our locale, that is. Irrigation, my dear water-freak neighbor, was invented to keep our fields of food alive, not your imported turf.

Get this: Of the estimated 7 billion gallons used each day for landscape irrigation, 50 percent is wasted due to runoff, evaporation, or simple overwatering. Our homes may be our castles, but we don't have to create moats to go along with them.

About half the world's population lives outside of a city. In fact, until 2008 most people lived in rural areas. In the United States, the majority of people live in the suburbs. That usually means they have a home, a yard, maybe even a garden. I'll spare you the conjured white picket fence. Still, chances are you live on that sort of a homestead. While you may not have tracts of land, a ranch, or a farm that siphons water, the way in which you manage your plot of land makes a difference—a big, big difference—to the total share of water available to your neighborhood, your town, your region, your state, and your country.

Some people have smartened up and try to do the right thing to lessen their effect. However, alternative approaches to watering are hindered in some cases by strange government policies and regulations. For example, using barrels to capture rainwater and then using that water for irrigation sounds like a neat and innovative way to help reduce demand on traditional sources, right? But it's illegal to capture rainwater in some states. Yup, as crazy as it sounds (and is), capturing rainwater isn't allowed in Colorado, Utah, or Washington.

Water rights in the western United States operate under a prior appropriation basis, otherwise known as first come, first own. If you were the first to lay claim to some tract of land, you can use as much water as you want from wells, rivers, streams, what have you, and then you own the right to that amount in perpetuity. Along comes the next person in line, and she can use as much as she wants of what's left, and so forth. In the eastern part of the country, riparian law holds. This system, based on English common law, states that an allotment of sorts should be used to make water available, as well as fair and accessible, to people whose properties are adjacent to the source; the rights of one homeowner or landowner are weighed against the rights of all.

This is where and how strange policy making interrupts individuals who are simply trying to do the right thing, something as simple as watering their lawns. And it's important because when we are out-of-doors, we are in the place where water gets used the most. Innovative water-saving actions should be encouraged, not discouraged. You, me, all of us can make a difference, given the right incentives and information.

So rather than stand there, hose in hand, without a clue as to what you can do, here are some water-saving ideas to help you better manage your share when you're outside.

EXTERIORS . . .

Gutter diverters send the rainwater to your landscape or into rain barrels for storage. Considering that the average home in a temperate region can satisfy its annual outdoor water needs with solely the amount of water that falls on its roof, it's silly to let rain go to waste. Every inch of rain that falls on a 2,000-square-foot roof totals more than 1,200 gallons.

Use your roof as a catchment area for water. Ever wonder what those giant silos on top of New York City high-rises are? They're water storage tanks. You can get smaller versions for your house. One tank owner in Texas, a very dry state, says he captures enough water to maintain his 2-acre lawn and garden all year long.

A "gray water" recapturing system recoups water that goes down the drain and diverts it into a storage tank to be used for irrigation. Some even have filters to rid the supply of soaps, toothpaste, and the like. You can use the water for your lawn or plants. Considering that just 25 percent of the water we use at home isn't recoupable (what's used for flushing and food waste), a couple of hundred gallons of wastewater could be recycled per day.

LAWN AND GARDEN AREAS...

Even though billions of gallons of water per year are devoted to our lawns, we can still cut use by caring for those areas properly. Speaking of cutting, 2 inches is about the best height for grass. Shorter grass requires more water. You can also leave the clippings as a sort of natural fertilizing blanket. So don't trash grass!

Get a sprinkler timer and set it for the early morning hours. Less water will be wasted due to evaporation compared to watering later in the day. You can waste as much as 30 percent of the water you use to evaporation by watering at midday. PS: Watering at night leaves grass damp, making it susceptible to disease. Best to water just before the sun rises.

How many times have you had to step off the sidewalk and into the street because some numbskull had his sprinkler pointed in the wrong direction? It's a common problem. The average sprinkler uses some 240 gallons of water per hour—all of it wasted if the beneficiary is the patio or sidewalk. Sure, the fix is hard (you have to turn the spout toward the grassy area!) and takes all of 2 seconds, but do us all a favor, please.

Use a shutoff nozzle on your garden hose. Water flows from an unrestricted hose at a rate of more than 12 gallons per minute. Twelve gallons is enough water to fill a nice-size aquarium—and there ain't no fish on your lawn (let's hope). Nozzle it.

If you have a Bellagio Hotel fixation and are set on having a fountain for your yard, make sure the radius of the basin is at least twice the size of the height of the water stream to best recapture spill. Also, affix a wind sensor and an evening timer so nothing gets blown away or wasted in the wee hours when the aesthetic is lost on all but the neighborhood raccoon. And it goes without saying that you're using a pump that recycles the water, right?

Get drought-resistant plants for your garden and landscape, and you could save up to two-thirds of the water you'd otherwise use on thirstier plant types. Some water utilities will even pay you to do it. One water district in California, for example, offers up to $2,200 to residents in its service area who replace grass with water-wise plants. That's a lot of green to do with less.

Even if you can't pronounce it, you can do it: xeriscape. *Xeriscaping* means instead of a grassy lawn, you use drought-tolerant plants, stone walkways, mulch, and dirt to create a landscape. An acre designed this way can save about 850,000 gallons of water annually, which is almost enough to supply six homes with their entire water needs for a year.

Watering your garden shouldn't be a daily activity. Once or twice a week is sufficient. Most outdoor plants are overwatered by 50 percent. Reduce water needs further by planting where there's shade. Grouping plants with similar water needs also makes a lot of water sense. Water slowly, too; watering fast causes runoff.

The type of soil you grow things in makes a huge difference in the amount of water your garden needs. Some soil is sandy and drains fast. Other soil has more clay in it, which poses drainage problems. A simple solution is to find out which you have by digging a hole 1 foot deep and filling it with water. If it drains within 3 hours, your soil is sandy and you'll need to water more (or plant more drought-resistant plants). If the water stands for more than 8 hours, you should water less to prevent oversaturation. Soil with adequate drainage should drain within 4 to 6 hours. Most gardens need about an inch of water per week.

The type of fertilizer you use can also reduce your garden's watering needs. Just 5 pounds of organic compost mixed into 100 pounds of soil (a 1- by 10-foot row tilled 6 inches deep) can hold an additional 25 gallons of water—which would otherwise drain directly out of the soil. Organic compost can be made from your lawn clippings and kitchen waste—tossed fruits, vegetables, even tea bags. Think of it this way: you'll be putting food (and water) back from whence it came.

AT THE POOL . . .

If you must have your own personal cement pond in the backyard, make sure it's water neutral. New pools can be designed to include tanks that collect rain and top off the pool as needed. They also utilize backwash minimization systems that prevent excessive spill. You can even retrofit a pool to water-neutral standards. In many cases, the rainwater tanks collect more water than is needed by the pool, so you can end up with surplus water for garden irrigation or other outdoor purposes.

Maintaining your pool properly is key to keeping its water use low. By keeping the water level lower, you reduce water loss due to splashing. You should also meter the water that refills your pool. If the volume you're using suddenly spikes—you have a leak. And use cartridge filters instead of sand filters. You don't need to backwash them to clean them. Clogs can cause overflow—and so can cannonballs! Keep splashing to a minimum to maximize water savings.

Don't let your pool lie there naked—cover it. Up to 95 percent of the pool water that is lost to evaporation could be saved by installing a pool cover. That isn't a modest amount. It adds up to billions of gallons lost every year across the country. There are some 9 million backyard pools in the United States, and yet only 30 percent have pool covers.

Another good reason, besides safety, to block off your pool with a fence is that the fence can act as a windbreak. Planting trees or shrubs nearby can serve the same purpose. A 7-mph wind at the surface of the pool can increase evaporation losses by 300 percent, which means you might need to refill your pool three times more often than average if you live in a windy area and don't protect your pool with a cover or windbreak.

You don't have to drain your pool every season. In fact, you really don't have to drain it for up to 3 years (or at all) if it's properly maintained. And no, this doesn't mean turning it into a skating rink in winter. You can use ice compensating technologies, such as winterizing chemical kits sold at pool stores, to keep your pool from freezing.

BOTTOM LINE

Don't overwater. Dousing the landscape literally with water won't make it any greener. In fact, it's one of the leading causes of floral death. Most outdoor plants, for example, are given 50 percent more water than they need.

Go natural. Indigenous plants and those that grow easily in the local area are what you should have. Nature usually provides sufficient water for its local habitat. Ban imports. It's called xeriscaping.

Maintain the drains. Keeping your gutters clear, your pipes pumping, and your pool clean prevents backups and spillage. Preventive maintenance keeps leaks and floods from ruining your home and the water supply. The average US home leaks 11,000 gallons a year.

3

work

make it a water recession

Water is big business. Just five beverage companies consume enough water over the course of a year to satisfy the daily water needs of every person on the planet. Of course, we may not be able to control how much water is put in a can of soda or a beer (less water, more alcohol, please) or the amount it takes to make paper, but we can control our own use at the workplace and even influence those who manage supplies.

It may not be our nickel that gets spent on the utility bill at work, but the gains are certainly ours when we reduce the corporate water footprint on the planet. Water prices are poised to rise due to increased water stress, and corporate growth is expected to be impeded as resources dwindle. Make no mistake, all of this comes out of our paychecks in one way or another.

Twenty percent of the world's water supply goes to support industry. That's twice as much as is used to support municipal supplies for our personal use. Sometimes we forget how many resources we use at the office itself. Many of us spend as much time at work as we do at home. Even if we aren't showering there (as philanderers and bicycle commuters often do), we can still use a lot of water throughout the day.

The average workday in the United States is 8.7 hours. We spend 1.1 hours a day drinking and eating, and up to 1.75 hours per week in the restroom (ladies spend just 1.4 hours). If we scale that back, per worker, by even a few cups per day on the front end and a few flushes on the back end, the savings would amount to more than 2 billion gallons of water. And that is about as much bottled water as Americans drink over the course of a whole year; profligate use, for sure.

Solutions abound in the workplace. A simple thing such as minding the amount of water used in the break room can add up to industrial-size savings. An extra cup of coffee left at the bottom of the coffeemaker each workday multiplied by every business in the country equals more water than a billion people need per day for drinking. Over the course of a year, manufacturing the paper we use at work requires 68 million trees and 82 billion gallons of water, or more than 1,000 gallons of water per worker. A meeting for a group of a dozen executives can easily utilize gallons more than it needs to simply by using bottles of water instead of a pitcher.

There are some 27 million businesses in the United States. Seeing as how we comprise those businesses, we can do more than earn our keep; we can save water at the office.

Business owners may also want to take note of the growing risks of water scarcity and climate change. There are already reduced water allotments, more stringent regulations, higher costs, growing community opposition to business use of water, and increased public scrutiny of corporate water practices. In other words, a true-life *Milagro Beanfield War* may be in the making.

"Businesses cannot afford to ignore this trend. For some it means new economic opportunities in making water available to meet demand or in finding solutions to improve water quality and water use efficiency. For others, it means closer scrutiny of how they, their supply chains, and their markets access and use water, and of how new business risks emerge as they compete with other users. In any case, it is time for businesses of all sectors and sizes to add water to their strategic thinking," the World Business Council for Sustainable Development said in its 2006 *Business in the World of Water* report.

Each of us, too, can add some strategic thinking to his or her business world and business life.

Here are some ideas.

IN THE BREAK ROOM ...

Mind the coffee in the break room. For every cup of coffee wasted, 590 cups of water are wasted, too. Brewing a "fresh pot" means using a lot of freshwater. There are about 27 million business establishments in the United States. A cup saved in every office keeps 16 billion cups of water from being tossed out, cold and unused.

Use a glass or a mug instead of a paper cup at the water bubbler when you drink and gossip about last night's reality TV show episode. It takes more than 6 gallons of water to make just one of those tiny 3-ounce cups.

IN THE RESTROOM ...

How many times have you found that someone has left the faucet running in the restroom? Custodians report that it happens as often as 10 times per week. That's 2 to 3 gallons wasted every minute. I'm all for employees washing their hands before leaving the restroom, but they could turn off the faucets. Infrared sensors overpower mindlessness; these sensors can be adapted to any faucet and save as much as 70 percent of the water used in the typical hand washing.

The type of faucet installed at the sink can save lots of water, too. High-efficiency faucets—those marked with the WaterSense label, deemed "water efficient" by the US Environmental Protection Agency—save 30 percent of the 600 gallons of water each person uses to wash his or her hands per year.

Dual flush toilets are not only water savers but also clog preventers. No more dance steps getting in and out of the stall. Of course, they do their main job well too, saving 67 percent of the water normally used by flushing.

In the men's room, waterless urinals make a lot of sense. Just one waterless urinal can save 45,000 gallons of water per year. A typical office building restroom with three urinals and 120 men could save almost 240,000 gallons of water per year. Expand that across the 8 million urinals installed in the United States, and you save as much water as 2 million people use in a year.

AT MEETINGS . . .

Use mugs and pitchers of water instead of plastic bottles. Mugs and cups allow people to drink as much or as little as they want. Ceramic coffee cups or mugs hold less than half the amount of water contained in the typical ½-liter bottles.

A 16-ounce cup or mug holds exactly that amount. Open a bottle and you own what's in it whether you finish it or not.

It takes 3 liters of water to make a single 1-liter plastic water bottle, and we use 98.6 million plastic water bottles each day. Put another way, that's 296 million liters of water, or the water equivalent of almost two bottles for every working person in the country, every day of the year.

Ordering too much food at the office just creates massive waste. The water it takes to grow and make food is exponentially more than the amount that's in the food when you eat it. Think about that the next time you are weighing your plate at the salad bar. Half of all the food purchased in the United States becomes garbage. Tossing out just a leftover piece of chicken can waste 115 gallons of water.

THE OFFICE LANDSCAPE . . .

Your office landscape should be managed just as efficiently as your home lawn. There are more than 705,000 office buildings in the United States, with their average area totaling 14,900 square feet. Even if the outside area of these buildings is less than half the size of the inside, office buildings are still responsible for more than 113,300 acres of landscape. It takes approximately 652,000 gallons of water to maintain an acre of land over the course of a year. Less green requiring less water can save billions of gallons a year—and a bundle of money in utility costs to boot.

How about installing a smart water system? What's that, you ask? It's an irrigation system that tracks weather reports before giving the okay for sprinklers to turn on. It helps cut water use by 59 percent and water runoff by 71 percent. One office building operator expects to save 42 million gallons of water a year with such a weather tracking system.

BOTTOM LINE

Mind your manners. Leaving the faucet running or letting the toilet overflow isn't something you'd do at home, and it should be the same at the office. Office mindlessness is more common than you might think (when it comes to water, anyway).

Don't drink and douse. Spending too much time at the water bubbler is one thing, but using a lot of cups is another. Use mugs and stop with the disposable cups and bottles—it takes a whole lot of water to make them, too.

Finish your food. Takeout doesn't have to mean tons of trash and garbage. Ordering six large pizzas for two people on the company's dime ends up wasting all the water that went into making those sausage slices sitting at the bottom of the Dumpster.

sports

keeping a low water score

I know what you're thinking: Don't bum me out with facts about how much water is wasted by my favorite sport. Sports are fun, or are supposed to be anyway. No need to take that joy away. But games can be played and watched without a lot of losing—losing water, that is. It's about creating a good water defense so the myriad offenses that sports wreak on our water systems don't win out.

Nearly 150 million Americans attend a baseball, basketball, hockey, or football game each year. Team and stadium owners, as well as league officials, are getting wise to different ways to save while still providing a great experience for fans.

The National Football League has a green advisory committee to help it, among others things, stop water waste. The PGA of America teamed up with Audubon International to conserve water and encourage wildlife preservation. And Major League Baseball hooked up with the Natural Resources Defense Council to create sustainable stadium operations and team practices.

Water savings is a top priority for sports professionals and enthusiasts because water is so critical to game playing, whether it's to keep the field green or the athletes hydrated.

Stadiums that typically use upwards of 40 million gallons of water a season are being built to reduce demand by 25 percent. Waterless urinals, synthetic turfs, and new parking lot medians reduce water consumption—and costs. The World Cup soccer organization and the International Olympic Committee have also set new environmental standards to cut water use and water waste. An international water-saving effort is taking place.

Still, many sports demand water in order to be played. There are those that require direct use, such as ice hockey, and those that require indirect use, such as golf. It may take 15,000 gallons of water to freeze a hockey rink, but the average golf course in the United States uses some 50 million gallons per year, and there are 23,000 courses in the country. In any event, you can easily see how water claims its place in the professional record books.

Now I promised not to bum you out. So here is the positive aspect of all this: Professional sports get it. They are water conscious and are cutting back where they can.

All of us amateurs can easily do our part, too. Almost 13 million people play softball recreationally. Then there are those of us who shoot hoops or are gym rats, yoga practitioners, runners, cyclists, or swimmers. Add us up and we're a significant portion of the population. This means we can make a real impact by trimming our water use while we try to stay trim and fit, or just have a little fun in our spare time.

As hacks, we can pretty much set the rules for the games we play and the activities we take part in. Nearly 40 million of us exercise on any given day. Yet about two-thirds of us choose to work out in the afternoon, adding an extra shower to our daily routine. There go 540 billion gallons of water. We can easily enough change our daily game plans. It means a little rescheduling and putting water on the playlist.

Just as professional athletes and managers develop winning strategies, so, too, can we as spectators and amateurs. We can dip, not splash. We can keep our water supply healthy for games, practice times, and training sessions to come. Now go out there and win one for the dipper!

Here are some game changers.

SUMMER SPORTS . . .

Swim in the center lane or middle of the pool to prevent excess splash. Splashing is one of the leading causes of pool water waste. If just one swimmer in every pool in the United States splashed just a half-cup of water, you could fill more than a dozen inground pools.

It takes about 3.1 billion gallons of water a day to maintain golf courses in the United States. Play in the morning or early evening if you can so you aren't treading on grass when it's driest and most susceptible to being torn up. If possible, choose a course that has implemented water-conservation design and maintenance principles. (Lots have, and the United States Golf Association recommends it.) The type of grass you play on matters, too: Buffalo grass, Bermuda grass, salt grass, and other low-water-using turfgrasses can minimize water needs by 50 percent or more. The extra 1.5 billion or so gallons saved per day if every golf course implemented these measures could supply the entire global population that lacks freshwater.

If you're a baseball fan, "keep it mellow" in the restroom. More people attend baseball games than any other sport. Okay, that's mostly because there are more games played in baseball than in any other sport. In any event, if every fan

who attended a game—and there are about 75 million of
you—kept flushing to a minimum by eliminating just one flush
each when they hit the restroom, we'd save a million gallons
for every day of the year.

**If you take a water bottle with you when you run (or pound one
after) and then toss it, don't.** Use a refillable container
instead. The extra 3 liters it takes to make a plastic water bottle
adds up to more than 100 million liters saved per run if every
jogger in the United States adopted this practice. Just do it.

Tennis anyone? Sure you can play anytime, but if you have clay
courts, you should water them only at night. Tennis pros say
eliminating a midday water cycle actually improves the
court's surface because drier areas "fight" for moisture from
wetter areas, creating a smoother mix all around. It takes
approximately 400,000 liters of water per year to maintain
just one clay court. Avoid a double water fault.

**We've all heard about the dangers of dehydration, but what
about overhydrating?** Hyponatremia may be even more
dangerous than dehydration. It means you drank too much
water and your sodium level has fallen too low. To keep
properly hydrated, it's recommended that you weigh yourself
before and after a workout. If you've lost weight, you're
dehydrated. If you've gained, you drank too much. Balance
out and save the precious water inside you. Sixty percent of
the human body, remember, is made of water.

Try working out in the morning. People who exercise in the
morning tend to be more likely to stick with their fitness
habit than those who sweat later in the day. Besides, working
out in the morning means you can skip that extra shower
you'd take in the evening après sweating, a savings of about
20 gallons. Ten percent of the population showers twice per

day. That's more than 600 billion gallons that could be saved every evening.

The type of surface you play on also affects the water system. Certain turfs and grasses require more water than others. And certain court types also tap the water supply, require hosing down, or cause runoff that harms other freshwater systems by polluting them. The more natural the grassy area, the better. New types of court tiles are now available that save on maintenance and runoff. These are water-resistant tiles that are unaffected by humidity and have small openings that allow water to pass through and drain away below the surface. No more puddles.

WINTER SPORTS . . .

Ice-skating takes water—no way around that. In fact, it takes about 15,000 gallons to make the average National Hockey League rink ready for play. The best ice for rink skating is $\frac{3}{4}$ inch thick and kept frozen at 16°F. Every $\frac{1}{4}$ inch of ice consumes 3,750 gallons. Try to skate or play hockey at off-hours, when fewer people are shaving ice off the top.

It can take 8 million liters of water per day to make snow for a typical ski area—and more than 100 times that for the full season. Many ski resorts are opting to use recycled water for their operations and have installed water-recapturing systems for the melt. You can check a ski area's environmental scorecard at www.skiareacitizens.com. Obviously, it's best to ski on natural snow and avoid undisturbed areas so you don't wreck natural barriers and force more artificial snow to be made. Ski when few want to: in season when it's crowded.

Use the drinking fountain inside at the gym. Fountains are usually fitted with carbon-based filters (and many also with chilling systems) that provide a continuous supply of freshwater. With 41.5 million health club members in the United States, saving the 3 liters of water it takes just to make one plastic water bottle adds up to almost 125 million liters that could be conserved per day. Besides, if your gym is anything like mine, more people are on their cell phones than working up a sweat, and the water bottle looks like it's more for show.

BOTTOM LINE

Where you play is just as important as how you play. The type of surface you play on affects the water system. Snow, ice, and pools put you in direct contact with water. But the ground beneath your feet also matters: the type of grass, clay, or soil you're hitting hard. So look down and check out if it's water tolerant.

When you play should be considered, too. A morning workout eliminates the need for that extra shower per day. And golfing at that time reduces the chances you'll tear up the green even if you're a high scorer.

What you play with has its water count as well. Taking a refillable water bottle with you whether you are watching or playing eliminates plastic waste and the extra 3 liters of water that go into making a liter-size bottle.

5

travel

away doesn't mean a waste

After fleeing the little people of Lilliput, Gulliver is forced back to land in search of freshwater. So go the wacky adventures in Jonathan Swift's *Gulliver's Travels*.

Few travelers today have to abandon their plans in search of freshwater. Indeed, traveling involves massive amounts of water consumption. Most of us in the United States take a vacation every year. More than 30 million Americans fly somewhere outside the country annually, and increasingly, we are opting to fly instead of driving domestically. And when we get to our destinations, we stay mostly in hotels or motels. It's there that we really let the water dogs out: The average luxury hotel room's estimated water use is 475 gallons per day, which amounts to more than the average US household uses! We turn into traveling, water-sucking giants who splash about merrily, drinking, bathing, steaming, basking in hot tubs, and waiting in line at aqua parks in the desert. Yes, we are a curious lot in different senses of the word.

Still, before we even leave our homes and begin acting water-crazed while on vacation, we waste loads of water in ways we may not even have thought of.

Leaving the hot water heater on when no one is there to enjoy the warmth could waste gallons. Leaving those sprinkler timers on when it rains and you aren't around to turn them off wastes even more. And spring a leak or have a pipe burst and look out—there's a flood of thousands of gallons. Pipes burst more often than you might expect, especially in colder climates. And the average household leaks 11,000 gallons of water a year anyway—enough to fill a backyard swimming pool. In fact, American homes in total leak more than 1 trillion gallons of water every year. That is the equivalent of the annual water use of Los Angeles, Miami, and Chicago combined. If you aren't around to spot a leak in your house, the problem only worsens.

Yet, we often don't just leave our water troubles behind; we take them with us. Bottled water seems almost a travel necessity for people. Too bad airport security doesn't feel the same way. Millions of liter bottles are confiscated each year, and the water in them is wasted along with the water used to make all the plastic.

Another mindless waste of water is the hotel laundry service. Every guest's laundry must be washed separately. (The laundry service isn't like your mom; it doesn't know which socks are yours.)

While we're at our travel destinations, we end up using, on average, three times more water than the people who actually live in the locale we're visiting. It's those long showers, hot baths, and excess, well, everything that we imagine we deserve when we are on vacation. And because we don't want to take risks with the local water, we use more of that bottled stuff.

For us visitors who find things alien to us "weird" and who somehow expect conditions no matter where we are to be just like at home, an education process needs to begin.

Here are some ideas that make a lot of water sense for your

home while you're away and lessen your water footprint when you travel.

LEAVING HOME . . .

Do you know where your main water valve shutoff switch is? If you are going away for even a weekend, it's a good idea to find it and shut it, especially in colder climates. Pipes burst more often than you might think: A major city can, for example, experience almost 4,000 pipe bursts a month in winter. In fact, about $33 billion is spent fixing burst pipes around the world per year. You can save the expense and headache, not to mention the thousands of gallons of water that are expelled, with a simple turn of the knob.

Okay, so if you aren't into turning off all the water to your home when you get the travel bug, then at least consider sprinkler timers with built-in rain sensors. Your lawn and garden don't have to be watered twice (once by nature and once by your sprinklers) just because you're away. Efficient home irrigation systems can save 11 billion gallons of water per year, which is like running 3,200 garden hoses 24/7 for 365 days.

Unless you're into keeping hot water on reserve for burglars while you are away, feel free to turn your water heater's thermostat down (some even have "vacation" settings). You'll save energy and the water needed to produce it, as well as money.

Overwatering your indoor plants before you go away or as soon as you get home isn't good plant care; it just wastes the extra water you pour into the pot. Plants can absorb only so much water at one time. Instead, try watering them just before you go and then wrapping your plants in those clear or light-colored leftover plastic bags you keep around the house (because you're using canvas shopping bags now, right?). The plastic seals in the humidity and keeps them moist and healthy.

AT THE AIRPORT . . .

More than 600 million people fly every year on US airlines alone. And lots of those people buy a bottle of water and chug it before going through airport security (where liquids are banned), only to buy another bottle after the security checkpoint. That's a giant waste and can add up to hundreds of millions of bottles—and gallons of water—wasted. So, bring your own reusable water bottle when you travel. It's one more thing to pack, but one less water worry to care about when you travel. PS: Don't forget to guzzle it before you go through security.

If you do buy a bottle of water at the airport, save the bottle (and the water that went into making it, at least). Security is only concerned about the liquid in the bottle, not the bottle itself. Hit one of the drinking fountains after the security check and refill it before you get on the plane. The Transportation Security Administration confiscates some 13 million items a year; don't let your water become one of them.

When you hit the airport bathroom . . . be quick about it. Airport stalls and urinals are typically equipped with timing sensors. If you stay in the stall for less than a minute, the toilet will flush a half-gallon less than if you stay in the stall longer than a minute. Assuming that every airline passenger goes to the toilet once, the savings of being fast could add up to 300 million gallons of water a year. Yet another reason not to sit around and tap your foot in the stall.

AT THE HOTEL . . .

Standard hotels use up to 200 gallons or more of water per occupied guest room per day. That's 73,000 gallons per year for one room! You may be on vacation, but there's no need to be a pig about it. The same water-conservation rules apply to your hotel stay—shorter showers, turning off the tap, etc.—as when you're at home. Fifty-six percent of the water used in a hotel room is for showering. It may seem like a vacation rite of passage to take a long, hot shower, but trimming back could save up to 50 gallons a day. With 4.4 million guest rooms in the United States, that's a lot of water to be saved.

That nice, ice-cold water bottle sitting in the mini bar is so tempting—yet so expensive. Don't give in to the temptation. Using your own water bottle versus that $7 bottle can save a lot. If just one out of every four hotel guests reused his or her own bottle instead of buying a new one, it could save more than 500,000 gallons of water a day (and lots of people from feeling like they just got ripped off).

Sick? Tired? Holed up in your hotel room and ordering room service? If you already have your own bottle of water, tell the in-room dining operator you don't want water with your meal. That extra glass saved may not seem like much, but if even just one person in every hotel in the country refused a glass, it would save more than a million gallons of water per year.

Did you know that a hotel has to wash each load of guest laundry separately because—think about it—they have no way of knowing whose is whose? That's a big waste of water. Because each hotel guest generates an average of 5.7 pounds of laundry every day, the water used to wash it adds up—to 1.2 billion gallons. Do yours when you get back home.

Ah, the spa. A commercial hot tub requires about 1,200 gallons of water, and a lot of that evaporates because of the heat. But the biggest drain on hot tubs is your stank. Sweat, soap residue, and oil clog tub filters, forcing the water to be drained more frequently (every couple of weeks). Be clean before you soak and don't spend more than 15 minutes in the tub. You'll help save more than 27,000 gallons of water per tub per year that way.

AT YOUR DESTINATION...

You're in a foreign land when thirst strikes. You shop for a bottle of water. Then you notice that the bottle is emblazoned with the same name of the place you are in. The water in many countries is of such good quality that they export bottles of it. It's simple enough to investigate the water quality of the place you're visiting to determine whether you're able to drink safely from the tap. More than 30 million US residents travel outside the country per year, the majority to countries with safe water supplies. A little research can go a long way toward saving water abroad.

Traipsing around places you shouldn't can damage the local water supply. Best to mind your feet and stay on designated pathways. Unduly disturbing dirt, especially in the dry season, can send dust into the air or directly into the water, or snow, or ice and pollute it. Increased activity in the Colorado mountains, for example, is causing more erosion and speeding up the runoff to rivers that supply millions of people with water. With 75 million hikers in the United States, it's an easy feat to cause a dustup.

How high are you gonna get while you are traveling? No, I'm not talking about drugs or alcohol. The elevation you will spend time at can determine how much water you'll need. You dehydrate more at higher altitudes—as much as a quart more than usual above 6,000 feet above sea level. Of course, if you stay nearer to sea level, you'll need less water. Sorry, skiers, but chalk one up for beachgoers!

A minus for summer travel is overall water use: It's peak season. Water use in summertime nearly doubles. Evaporation is stronger. Moreover, vegetation is flourishing, putting demand for freshwater at its highest all year. This strains supplies. If you are invading—I mean visiting—a summer vacation spot, be aware of your use, splashes, and spills. You may even want to consider off-season travel. It will cost you, and drain the place you are visiting, less. If, for example, people visiting beach-famous California trimmed their summer water use (along with all of those wasteful permanent residents), it could shore up supplies to help prevent the state from running out of water in the next few years.

BOTTOM LINE

Locking up should leave little room for error. Imagine if you weren't around to catch a bad leak or a pipe bursting?

Shutting off the main water valve to your house makes a lot of sense and may help prevent a call to your insurance company.

Your hotel may not charge you (yet) for the water you use in your room. But that's no reason to use more water than you would at home. And certainly having your hotel do your laundry is a poor (and expensive) water option. Be a good guest and watch your water use.

It may be okay to drink the water. Many countries have cleaner water supplies than you'd find at home. No reason to drink from the bottle when you're visiting a place whose name you can find on the label of bottled water at the supermarket.

section
2

the water you can't see

a product guide for making smarter choices

6

foods and beverages

how much does it take to produce your produce?

"Excuse me, bartender. The beer you just served me had 20 gallons of water in it." You might be thought drunk if you said that, but it's true.

The water we are talking about here is virtual, or embedded, water. It's calculated by totaling all the water it takes to grow, raise, or manufacture something; it's the water we don't see in all the things we drink, eat, wear, and use in our lives. Turns out that this unseen water drains our supplies more than the water that's right before our eyes. Think of it this way: When you toss out that cup of cold coffee in the bottom of the pot, you're actually tossing out 590 cups; it takes that much water (37 gallons, to put it another way) to grow the coffee beans needed for just that 1 cup.

Most of the water that you "drink" actually comes hidden in the food you buy. Agriculture is the number one consumer of freshwater in the world, accounting for about 70 percent of its use. And most of the crops we raise, we eat.

Our biggest effect on the world's water supply, therefore, can be waged in making dietary choices. Please note that I'm not saying you have to disavow the foods you like—we still have lots of choices. I'm just pointing out that by choosing what we eat a little more wisely, we can lessen the demand for some water-intensive foods and help shore up water supplies. For example, swap a hamburger for a veggie burger just once, and you will save about 750 gallons of water. A simple substitution here and there can add up to a whole lot of water savings.

This is why it helps to know how much water is in what.

Being smart about where we grow things and creating demand for foods that require less irrigation are ways to save water—and save lives. And they can save in more ways than just those: What's good for water conservation typically is also good for other ecological issues—as well as economic issues. While we are lowering our virtual water footprints, we also are lowering our carbon footprints. So it's important to learn who, what, where, and how something is grown or produced. This type of knowledge can also help us make healthier choices about what we buy.

This is how the trade flow of virtual water can help the overall water flow: a country with a lot of water grows water-intensive crops, which countries with scarce water supplies then import, allowing them to save their own water supplies.

An arid country such as Jordan in the Middle East saves as much as 90 percent of its domestic water supply by importing water-intensive products. Many Third World countries deplete almost their entire water supplies just trying to grow food, making disease and dying of thirst a daily reality for too many people.

Better management of virtual water can help. Virtual water management has already saved 5 percent of the water used in agricultural production. And now that the USDA requires most foods to have labels listing their country of origin, it's easier to make that water-smart choice by picking a water-intensive product that was produced in a water-rich region.

But being smart isn't just about the food we purchase, it's also about the food we waste.

Up to half of all the food we grow never hits our plates (or our bellies). Now think about all the water that went into that food's harvesting, production, processing, transportation, and storage. Add to that all the food that we toss out—food is the third most common type of refuse found in landfills, according to the US Environmental Protection Agency—and we have a lot of wasted water. Although we're talking about virtual water, the ramifications are just as real as if we were to let the tub overflow.

Buying the right foods in the right-size portions can trim our water diet plenty, to say nothing of the health benefits that will likely ensue for both ourselves and the planet.

Here's your shopping list.

FRUITS . . .

Apples—18.5 gallons per apple. Apples are grown in every US state, so choose those grown closest to home and save the water it takes to ship other varieties. By the way, an apple a day adds up to almost 7,000 gallons of water per year. You can keep the doctor away and save water at the same time by visiting your local farmers' market.

Apricots—19.8 gallons per serving, which is about 3 fresh apricots. A serving of dried apricots is 6 pieces (they shrink to half their size) and takes about the same amount of water. If you want fresh apricots, their growing seasons are mid-June through July in California and mid-July through mid-August in Washington, so try to buy local and in season where possible. Keep the dried ones for the winter.

Avocados—42.6 gallons per avocado. The Hass variety of this fruit (yes, an avocado is a fruit!) is grown year-round in California, which produces 90 percent of the nation's avocados. But due to a 3-year-long drought, try to avoid purchasing produce from this region during the drier summer months, when water resources are especially stressed.

Bananas—17.5 gallons each. The top-selling fruit in the United States is almost 100 percent imported. Bananas aren't ideal in terms of ecological impact, but the closer to home and the more organically they're grown, the better. Did you know that some say it wasn't an apple that Adam and Eve allegedly munched on, it was a banana?

Grapefruit—16.4 gallons per pound. This sweet 'n' sour fruit comes in white, pink, and red varieties, all of which can be harvested around the same time (between October and April). Grapefruit do best in warm, humid climates—trees can produce fruit 6 months sooner in warmer areas—and that means less watering. It also means thinner peels, which makes eating them easier, too.

Kiwifruit—15.4 gallons each. Despite the name, the biggest exporter of kiwis is Italy. You can find US-grown ones if you look really, really hard, but it's worth the effort—the closer to home, the more water saved. Available from late summer into fall, the sweeter ones are available in November. FYI, the biggest "Kiwis" are found in New Zealand (that's the colloquial term for local people) and—in the man and woman varieties—are about 5 feet 6 inches tall!

Lemons—4.8 gallons each. A cross between a lemon and a tangerine, the Meyer lemon is the heartiest of the lemon bunch, with trees that live longer, and it can be found in the United States (mostly in California). The Eureka! (exclamation point mine) is the most common, however—again, look for California grown. Lemons are largely grown in winter months and harvested in spring. Get 'em in season and pucker up.

Limes—4.8 gallons each. The peak season for limes is May (just in time for Cinco de Mayo!) through October, so get them then. And there are really only two types: key limes, like those from Mexico, and Tahiti limes, which are bigger. Get the smaller version because they are grown closer to home, and keep them out of the sun and bright light so they keep longer. PS: Don't slice and waste. Keep extra cut limes in the crisper in your refrigerator. They can stay good for weeks.

Mangoes—81.9 gallons each. Mangoes are generally grown on drought-resistant trees, but they fruit better when they're watered somewhat during dry seasons. Most mangoes are imported into the United States, but Florida and Hawaii grow some varieties commercially. Because of their relatively higher water footprint compared with other fruits, try to save your mango fix for when you're in the tropics, where nature does the irrigation for you.

Melons—15.3 gallons per pound. Melons come in different shapes and sizes and can be called fruits or vegetables, depending on who's asking. (What we in the United States commonly call a cantaloupe is actually a muskmelon; cantaloupe is a specific French variety. *Sacre bleu!*) Make sure you're ready to munch when you cut up the melon, though. Once sliced open, it's only good to eat for up to

4 hours (unless refrigerated), then you have to toss it—and all the water that went into growing it.

Oranges—13.2 gallons each. Avoid those grown in colder climates and seasons, because farmers spray the trees with water to create an ice coating that protects the fruit from further freezing. This may save the trees, but it sacrifices a lot of extra water.

Peaches and nectarines—11.2 gallons each. A summertime favorite with a medium-size (in terms of fruits) water footprint. Canning peaches for use in winter uses water, from boiling the jars, to making the syrup, to soaking the peaches. However, buying them imported from South America off-season isn't a great alternative, either. So, try to keep those peach cobbler cravings to the summer months when you can indulge in fresh and local.

Pears—7.8 gallons each. Pears are hearty and can withstand cold, but it's best to get them in autumn and let them ripen at home (sticking them in a brown paper bag can help). They don't ripen on the vine, so there's no need to sit around and wait. China is the world's biggest producer (with the famous Asian pear), but the United States produces a lot, too (the Bartlett, for one). Get the domestic kind.

Pepper (black)—589.7 gallons per pound. Black pepper is a vine that is grown for its fruit, peppercorns, once dried. There are black, white, and red types, which are made from the same fruits using different processing techniques, white being soaked in water for about a week. Black pepper is also made by soaking the peppercorns in water, but for less time. Get it dark, but look out: too much and you'll be sneezin'!

Pineapples—34.5 gallons per pineapple. Pineapples are one of the few sources of the enzyme bromelain, which has been shown to aid in everything from digestion to fighting respiratory infections. So eat your pineapples, but don't drink 'em. Pineapple juice requires about 361.2 gallons of water for every gallon of juice. For a guilt-free piña colada, use crushed pineapple instead. De-e-elicious.

Plums—14.3 gallons each. Plums have a short season and are at peak availability in June, July, and August. Get 'em fresh and never store them in the fridge. Besides the extra water you'd waste keeping them cool, they are sensitive fruits that experience "internal breakdowns" below 51°F. So keep your plums good and sane at room temperature.

NUTS ...

Almonds—259.2 gallons per cup of whole almonds (fresh or dried). This is a healthy, low-water alternative protein source compared with beef or pork. All US almonds are grown in California. Try buying them in bulk to reduce packaging— almonds last a long time when stored in an airtight container.

Coconuts—320.6 gallons per coconut. Coconut production consumes 2 percent of the water used for all crops. Buy them whole and use the water inside, too—just like Gilligan!

BERRIES ...

Blueberries—13.8 gallons per cup. Blueberries are indeed blue, requiring almost four times as much water per cup as strawberries. They do, however, offer great health benefits, so rather than eschewing them altogether, reduce their water footprint by washing them in a container rather than under running water, and perhaps make that next batch of blueberry pancakes a strawberry-blueberry combo: they are more colorful, and less water required. Blueberries, like strawberries, also freeze well, so you can avoid buying foreign imports by stocking up and freezing them in the summertime.

Grapes—14.8 gallons per bunch. White grapes mature quicker than red grapes and therefore require less water to grow. As we each, on average, eat about 8 pounds of grapes a year, that adds up to 237 gallons of water. Every grape counts. From the more than 60 varieties of grapes, choose local (the most popular grapes bought in the United States—Thompson seedless—are grown here anyway).

Raisins—44 gallons per cup; grapes—about 10 gallons per cup. It's best to eat 'em fresh: Raisins aren't as dry as they look. A cup of them takes more than four times as much water as the grapes from whence they came (being smaller, you've got to pack more in to get the same amount).

Raspberries—18.4 gallons per cup. These have the highest water footprint of the four berries we list here. Grown commercially in the Pacific Northwest and California, fresh raspberries are found throughout the country during summer. So, if you've got to have 'em, at least buy locally and in season to reduce water wasted on packaging.

Strawberries—3.6 gallons per cup. In the summertime, strawberries are widely grown in states across the country, so purchase locally. They're also easy to grow in pots, so you can grow your own even without a garden. Gently rinse fresh strawberries just prior to eating (rinsing removes the berries' natural protective coating, making them more likely to go bad). Craving strawberries out of season? This summer, freeze a bunch from your local market and, come fall and winter, substitute frozen strawberries for fresh ones in recipes. Just don't thaw them completely or they will become mushy.

VEGETABLES . . .

Alfalfa (sprouts)—14.8 gallons per pound. The United States is the world's largest producer (alfalfa hay is often used for livestock feed). But get it in spring and summer, before it's only grown in water-stressed areas such as Arizona and southern California.

Broccoli—27.4 gallons per pound. Broccoli likes cool weather, and its peak season is between October and March. Get it from California, where most of it is grown, or close to home rather than from another country. And sure, we like to steam it, but save some water and try it raw.

Cabbage—20.8 gallons per head. The smaller the head, the less water it took to grow. Cabbage is harvested in fall and early winter, and "winter cabbage" (savoy cabbage) stays in season through spring. That's the most commonly grown variety in the United States, and not a bad choice. You don't have to boil it all the time, either; shred it for coleslaw to save the water you'd use in the pot (and the house from the stink!).

Carrots—6.5 gallons per pound. While "Bugs" may prefer the big ones to chew on, the smaller, the better for us, because they take up less soil and need less water to grow.

Celery—6.5 gallons per pound. Why does celery always seem like it's wet? Because it holds a lot of water, and it's grown in wet, humid places. Some of the most nutritious parts of celery often go to waste: its leaves, roots, and seeds. Use it all for soups. Many farmers' markets carry celery. If not, look for the common Pascal kind, which is mostly grown in North America.

Corn—108.1 gallons per pound on a global average. Eight percent of all global water used for crops goes to corn. Most of it is grown in the United States (50 percent of world production). For the best water savings, get unshucked ears grown in the USA, preferably in the North and Midwest, where the varieties grown require less water.

Cucumbers—28.4 gallons per pound. They're the easiest vegetable to grow yourself. Get them fresh in summer—not pickled in water.

Garlic—0.21 gallon per clove. Choose loose garlic bulbs—with white, papery skins—where possible to avoid the wasteful packaging. Besides keeping away those pesky vampires, garlic is a powerful antioxidant with many health benefits. And with its relatively low water footprint, you can use it liberally to boost your health and the flavor of your favorite pasta dishes.

Lettuce—10.43 gallons per pound. The most popular lettuce is iceberg. It got its name from the beds of ice used to ship it east. That should tell you something: It's hard to store. Romaine is easier to grow, less sensitive to heat, and has a longer shelf life. It also has more nutritional value (iceberg is 95 percent water). Besides, it tastes better.

Mushrooms—indirectly, several gallons per pound. Unlike plants that require watering, mushrooms are parts of a fungus (which grows off other things) and need only moist, nutrient-rich soil to grow. As long as humidity levels are around 95 percent and soil moisture levels are around 75 percent, they will thrive. The most common mushroom is the

Agaricus bisporus, that white, button-looking type. They are high in nutritional value and have medicinal value, and many people think they can save the world because they tax natural resources so little and can be grown in such abundance.

Onions—25.6 gallons per pound. There are red, yellow, and white onions, as well as shallots and many other varieties. But the way to save water is to not cry. By slicing or chopping onions in a basin of water, you'll avoid releasing the gas that sets off all those tears. The best water-saving advice, however, is much simpler: Don't cut the root of the onion. This lets off the most gas. Also, the finer the knife, the less cell damage and the less gas produced.

Peas—10.2 gallons per cup of fresh peas. Because peas freeze so well, most are processed (canned or frozen). But frozen peas take almost 50 percent more water (14.5 gallons per cup) and can't compare in taste to fresh peas. Peas are grown year-round in states across the country. See if you can find some fresh pods near you.

Peppers (bell)—18.1 gallons per pound. Unlike their fiery cousins, these peppers can be quite sweet, but it comes at a price: more water makes a sweeter pepper. But so does time—sugars and nutrients accumulate as the peppers ripen. So to save water, go for the red ones that have sweetened due to time, rather than to extra water.

Potatoes—12.7 gallons per pound. Several types of russets are fairly drought resistant. And those grown in south-central Idaho require even less water than those from the southwestern part of the state.

Spinach—12.3 gallons per pound. Don't get it prewashed. And don't get it in the bag or the can, even if Popeye says so. It takes water to make packaging.

Squash—40.7 gallons per pound. Squash varies in size from small zucchinis to giant pumpkins. And there are a lot of different types: butternut, acorn, and spaghetti, for example, as well as colors: yellow, green, orange. In general it's best to get summer squash when it's harvested and eat it almost immediately with little or no cooking. Winter squash, on the other hand, is harvested at summer's end, cured, and then stored—all of which requires water. The word squash actually comes from the American Indian word *askútasquash,* which means "a green thing eaten raw." Try to keep true to its name.

Sugar—100.4 gallons per pound, or about a gallon per teaspoon of raw cane sugar. Sugar made from beets (grown in temperate environments in North America and Europe, mostly) takes less water, about 71.7 gallons a pound. If possible, try to find raw beet sugar. Even better, switch to honey—raw and from bees frequenting local wildflowers or berries—which takes very little water.

Tomatoes—1.3 gallons per 2.5-ounce tomato. Africa and Asia tend to have higher water use per tomato, so it's best to stick close to home and buy in season. If you're shopping at a local farmers' market, buy from someone using drip or trickle irrigation—this technique often saves water, provides the consistent watering that tomatoes need, and avoids wetting the leaves, which can lead to disease. How will you know if the farmer uses this technique? Ask!

MEATS...

Beef—1,581 gallons per pound. Boneless cuts tend to be higher in water content than cuts with the bone (1,122 gallons per pound). Either way, beef is the biggest consumer of water of any meat. How 'bout swapping a hamburger for a veggie burger? Better yet, make it a salad. Making the trade just once will save about 750 gallons of water.

Chicken—468.3 gallons per pound as a global average. Most of the water footprint of chickens comes from the grain they eat. Consider substituting turkey, which takes about 200 gallons per pound less water than chicken. If you do buy chicken, go for deli-sliced meat to avoid packaging, organic to avoid hormones, and free-range or Certified Humane Raised and Handled to support humane treatment of the animals (e.g., no beak cutting or forced molting).

Lamb—398.8 gallons per pound. It's lamb meat when the sheep are young and mutton when they get old. Lamb requires less feed and water (because lambs are smaller). But since US sheep numbers have fallen off ba-a-adly (you knew that was coming) because of drought in the Southwest, where they are largely raised, you may want to wake up to that and stop counting sheep as part of your dinner menu.

Pork—614.3 to 648.0 gallons per pound, depending on the cut. Cured meats run a little higher (676.3 gallons per pound), and sausage and similar meat products (made with pork or a combination of other meats) are almost twice as high (1,176.7 gallons per pound). Pork uses less water than beef, but more than other protein sources such as turkey, chicken, eggs, or beans. Try to limit those late-night salami runs and buy fresh from the deli to limit water used in packaging (as well as nitrates and other preservatives that aren't good for you). Wilbur, your heart, and the planet will thank you.

Turkey—286.3 gallons per pound, or about 72 gallons for a good turkey sandwich. As far as water requirements go, turkey is one of the best meats. Try to buy from the deli counter to avoid excessive packaging, or perhaps even support local farms that provide a free-range habitat and humane treatment.

DAIRY . . .

Butter—3,602.3 gallons per pound. Holy cow, that's a lot of water. You have to take into account the feed for the animal to make the butter, plus all the processing. It's better hard and salty than soft and bland. Soft butter holds more water. Meanwhile, salt repels water granules in the churning process.

Cheeses—about 414.2 gallons per pound. For those of you watching calories and cholesterol, fresh cheeses (such as ricotta, cottage cheese, and cream cheese) require only 260.5 gallons of water per pound, about half as much as other kinds of cheese. They're also better for your heart and waistline and the planet—now that's something to smile about. Say cheese.

Eggs—22.8 gallons per large egg, or about 60 gallons for a plain omelette. Eggs are a better source of protein than meat in terms of water requirements, but watch out for cholesterol. The major egg-producing states are Iowa, Ohio, Pennsylvania, Indiana, and Texas, but many small farms raise chickens, so buy local whenever possible.

Milk—about 720.1 gallons per gallon of milk (that's an average for fat-free and low-fat). Whole milk and cream take almost twice as much water (nearly 1,317.0 gallons of H_2O per gallon of milk) as fat-free or low fat (less than 1 to 6 percent fat). So, if you've got milk, get the fat-free kind. And don't be tempted by the powdered stuff—it requires almost 4.5 times as much water as liquid milk!

Yogurt—about 36.3 gallons per 6-ounce container. This is a good, low-water calcium source compared with cheese and milk. Plus, choose the "live and active culture" (also known as probiotic) varieties, and you'll get the added benefit of "good bacteria" that can help with everything from digestion to immune system function. These bacteria also make yogurt digestible by lactose-intolerant people.

GRAINS . . .

Barley—about 100 to 200 gallons per pound, depending on how it is processed. Almost half of the barley produced in the United States (it's the fourth largest grain crop) is used for animal feed. So, again, reducing consumption of meat will cut down on water costs.

Oats—122.7 gallons per pound. It's a bit more (193.8) for rolled or flaked oats, or 10.3 gallons per bowl of oatmeal, plus a cup of water for cooking. Oatmeal provides a low-fat, high-fiber, protein-rich option that won't slow you down. (Baked products, such as bagels, give you those carbohydrate crashes.) Avoid adding milk or cream to your oats, and stick with water as the base. You can flavor oats with low-water spices like nutmeg. Next breakfast, try swapping out the bacon for a side of oatmeal with your eggs—you'll use less water and give your body a healthy boost to start the day.

Rice—about 200 gallons per pound, or 96 gallons per cup.
Rice accounts for more than 20 percent of the global water
consumption for crop production. The least-water-intensive
rice grows in regions with high rainfall and humidity that
sustain rice paddies without much additional irrigation.
Jasmine rice from Thailand is one good option. In the United
States, long-grain rice grown mostly in the South is a better
bet than short- and medium-grain rices, which are grown
primarily in California and Arkansas, where farmers rely
almost exclusively on irrigation.

**Rye—39.8 gallons per pound, or about 8.8 gallons per cup of
rye flour.** It takes about 2 to 3 cups of flour to make a loaf, so
your average pumpernickel costs less water per loaf than
wheat bread. When possible, mix in some rye with your wheat
flour and your loaf will be far less soggy.

**Wheat or white flour—approximately 101.7 gallons per pound,
which is the same as a loaf of bread.** Wheat makes up
almost 12 percent of global water consumption for crops. Per
calorie, wheat requires far less water than rice and is
therefore a good alternative carbohydrate. Avoid bleached,
enriched white flour (which makes spongy white bread). The
more processed, the more water—and the less nutritious.

CEREALS...

**Cornflakes—47.7 gallons per 18-ounce box, or 2.6 gallons
per bowl.** But that's just for the corn part. Then, add in the
water used for the other ingredients (sugar will run another
1½ gallons per bowl, and malt and corn syrup are the next
biggest components) and for processing (the corn grits are
cooked in steam pressure cookers) and, of course, the water

for the milk you add (45 gallons there), and that bowl of cornflakes can be well over 50 gallons a bowl—more than three times the water footprint of oatmeal! If you must have your cornflakes, try to find organic, no-sugar-added kinds to reduce the water wastage and save on calories.

Flax—375.5 gallons per pound. Flax comes from a versatile plant used to make linen, linseed oil (used for industrial products), flaxseed oil, and straight-up flaxseeds. Both the seed and the oil are excellent alternatives to fish as sources of omega-3 fatty acids. Given the current declines in many fish species' populations and the relatively low water footprint of flaxseed, you'd be wise to skip the farmed fish and hit the farmed flax instead. The best way to consume flax is as ground seeds, to get the fiber and essential fatty acids. It's also cheap. You only need 2 to 4 tablespoons a day, and at $2 a pound, it's way less expensive than salmon! Sprinkle 'em on everything from salads to oatmeal.

Granola—about 65 gallons per cup, but highly variable depending on the ingredients. Word to the water wise: Make your own. It's easy, and it avoids the water associated with processing (especially all the sweeteners that tend to be added to premade granolas) as well as all the packaging. When made using rolled oats, almonds, raisins, dried apricots, and honey, a homemade granola tastes oh so good.

Wheat bran—just more than 100 gallons per pound. Other common sources of bran are rice and oats, but they have bigger footprints. Bran is an excellent source of fiber, protein, and minerals. Bran is the seed coat, which is actually removed when grains are refined, such as when turning whole wheat into white flour or brown rice into white rice. So, stick with whole wheat, and you'll get the bran (and all its goodness) included in the footprint of the grain.

PIZZA AND PASTAS ...

Pasta—230.5 gallons per pound. A pound of pasta will serve
about six people. To reduce the water footprint of your pasta
meal, look for recipes such as pasta primavera that use
vegetables with low water footprints, such as peppers,
onions, garlic, and spinach. Although tasty, those meatballs
and cheesy sauces add a lot of water to the dish. And
remember, you only need enough water to cover the noodles
(about 1½ quarts for a pound of pasta).

**Pizza—312 gallons for a margherita pizza for two (10-inch
diameter).** Unless fresh tomatoes are available nearby (i.e.,
they're in season where you live), go for pizzas that use
tomato puree sauce over the fresh slices. Most of the water
footprint is from the mozzarella, but adding any toppings,
especially meats, will raise the water content rapidly. Stick to
the traditional and savor the water savings.

BEANS ...

Beans—56.2 gallons per pound. There are, obviously, many
different types of beans. But they can be classified by how
they are grown: along a pole (such as green beans) or on a
bush (such as lima and kidney beans). Bush-grown beans are
generally considered the easiest to grow and require the
least fertilizer and water because they have a short growing
season. Try Indian Woman Yellow beans or Jacob's Cattle
beans. You can silence "musical beans" by soaking them in
water to eliminate the sugar that leads to flatulence. But
since this is a book on water saving, I choose to be musically
inclined.

Soybeans—224 gallons per pound. There are so many soy
products these days that it's sometimes difficult to figure out
what's real and what's imitation. There are soy hot dogs, soy
bacon, soy milk, and many types of soy meats. The less
processed they are, the better.

BEVERAGES . . .

Aluminum cans—1.1 gallons per 12-ounce can. Of course,
there is also the virtual water of whatever is inside, but you
get water savings with recycling. Because 90 percent of the
virgin materials are saved when a used can is recycled into a
new one, recycling creates a further 40 percent in water
savings in the manufacturing process and cuts down on
water pollution by 76 percent, because water isn't fouled in
the mining process. If it isn't obvious: Recycle and make sure
your beverage manufacturer recycles, too.

Glass bottles—1.1 gallons per 12-ounce bottle. Glass bottles
are recycled at a better rate (about a third of those
produced are recycled) than plastic PET (polyethylene
terephthalate) bottles (less than a quarter of which are
recycled), so they are a better choice. The bigger-is-better
motto also holds true with glass. Besides, I think things
bottled in glass taste better.

1-liter plastic bottles—3 liters per bottle. That means we
waste 47 million gallons of water per day, because we toss
out 60 million plastic water bottles in the United States every
24 hours. That is the tragic irony behind bottled water. Drink
water from the tap, not a plastic bottle. For other drink types,
get bigger sizes. At least you'll get more liquid per container.

2-liter plastic bottles—132 gallons per soda bottle. While the bottle itself takes a couple of gallons to make, it's the sugar and all the other ingredients inside that suck up the rest. Less sugar equals less water. In other words, drinking diet soda puts your water consumption on a diet, too!

Coffee—37 gallons per cup. We drink a lot of coffee: Making it requires more water than producing any livestock or crop. After oil, it's the most traded thing in the world. To make it good to the last drop, get fair-trade-certified coffee. Your buzz will have come from a plantation that practices sustainable, socially responsible coffee farming, and that means better quality and more responsible use of water. Eye-opening.

Tea—5.5 gallons per cup of black tea. Most tea comes from rainy areas, except the types grown at lower altitudes, where irrigation is used. So, go for green and Darjeeling teas, which are grown in the mountains high; their flavors are also better! Take that with your crumpet.

JUICES...

Apple juice—349.2 gallons per gallon. The same rule applies for all juices: Get 100 percent juice. Sugar additives and preservatives are common in apple juice, but they contribute to more water being used to make the juice. Apples are tasty enough on their own.

Orange juice—272.2 gallons per gallon. Get it fresh squeezed. "From concentrate" means it had water removed and had, or will have to have, water put back in again. Too much water work—and waste.

Pineapple juice—361.2 gallons per gallon. Since squeezing a pineapple isn't so easy (try it, I dare you), go for 100 percent pure. At least you'll know that all its nutrients are in it.

Tomato juice—87 gallons per gallon—by far the least virtual water of the common fruit juice varieties. Again, the 100 percent pure or 100 percent juice kinds are best. If it's labeled "cocktail," "drink," or "punch," it's been diluted (with more water) and sweeteners have typically been added.

SNACKS...

Candy—7.4 or more gallons per pound. The most prevalent ingredient in most candies is sugar, which accounts for as much as 60 percent of the product's volume. Try licorices instead. Licorice is 50 times as sweet as sugar, so you need less to get your fix. Also, licorice actually improves water tables and helps other crops grow because it's a soil enhancer. Now, isn't that sweet.

Popcorn—9.5 gallons per 1.4-ounce bag of kernels. Pop it yourself. The processing, additives, and preservatives that go into prepopped bags require tens of gallons of water. Don't add salt or butter, and you'll save even more. The microwave stuff may seem more efficient, but the plastic wrap, paper bag, and gunky coating of oil or butter inside increases the water content. Make it the old-fashioned way: in a pot, and shake it well.

Potato chips—48.9 gallons per 200-gram bag. Get them unflavored and with the fewest additives possible. Sure, those sour cream and onion ones taste good, but the more spices used, the more water it takes to grow them. They also make you thirstier!

WINES AND SPIRITS . . .

Beer—19.8 gallons per 8-ounce glass. Most of the water that beer requires goes into growing the barley. Hoist a locally made microbrew and avoid the water that goes into transportation, preservatives, and storage.

Gin—114.5 gallons per liter bottle. Gin is made by adding juniper berries and sometimes other natural ingredients to grain alcohol. Try compound gin, which is made more like vodka instead of distilled gin, which involves "redistilling." The fewer processes used, the less water used, even though aficionados may take issue.

Red wine—31.7 gallons per 4-ounce glass. The younger the wine, the less time (and water) it takes from vine to mouth. Try younger varieties such as Beaujolais. And look for wineries that "dry farm," grow their grapes with only sunshine and rain. No irrigation required.

Rum—33 gallons per liter bottle. Unlike other spirits, there is no set process for rum. It's basically molasses and water fermented. But generally speaking, the higher the alcohol content and the lighter the color, the better—for water count, anyway. Higher alcohol content means less dilution. And makers often darken rum with caramel, which, again, means additional water is used. Pirates, by the way, used to drink grog, which is a mixture of rum and water. Arrgh!

Tequila—64.7 gallons per liter bottle. Avoid the mixtos. They are made with less than 100 percent agave and often have water-intensive additives and coloring—not to mention the makings for a nasty hangover.

Vodka—79.8 gallons per liter bottle. "Vodka" is literally translated as "dear little water," which should give you some idea of what's in it. Water and alcohol are the only two ingredients in pure vodka. Leave the flavored kind behind the bar. Those types have more ingredients that take more water to grow and infuse.

Whiskey—430 gallons per liter bottle. Different grains and processes make for different whiskeys (bourbon, for instance, is made from corn), and they're further characterized by where they are from (Ireland, Scotland, etc.). Single malts are best because they come from a single distillery and use only one grain (which means they use just enough water to grow that one). They are "purer."

White wine—28.5 gallons per 4-ounce glass. The grapes in white wines, in general, take less time to grow than those in the red varieties, saving on water. However, some varieties do take longer than the grapes in some early season reds. Try Chardonnay. It's an early budding vine and typically a less finicky grape. Besides, you can get it in a big bottle or even a box. The bigger the container, the more of what you want there is inside—and the less packaging waste.

BOTTOM LINE

Choose food that's grown over food that's raised. Animals require a lot more water than fruits or vegetables. Besides the water that animals drink, it takes water to grow their feed and process their meat. Take a step down on the food chain and save.

Buy local (such as at food coops and farm stands). It isn't just the water saved in transportation and storage that makes such a big water impact, it's also more likely that locally grown food will be produced more sustainably, with fewer pesticides and pollutants that can harm water supplies.

Get your fruits and vegetables in season. Your taste buds will thank you, and you'll help create less demand for goods that need to be stored or shipped from far away. When it's summer in Brazil, it's winter in the United States, remember. Peaches 'n' cream shouldn't be served at Christmas dinner in Maine.

clothing

are you wearing water?

"Fast fashion" describes the rapid life cycle of clothing from sales rack to landfill—and it's getting faster.

Just like fast food, fast fashion is momentarily good and then gone, leaving you feeling pretty much unfulfilled and fat a lot of the time. But fast fashion leaves behind more than just an empty carton, a plastic cup, a hamburger wrapper. According to the US Environmental Protection Agency (EPA), more than 8 million tons of clothing and footwear enter the waste stream annually, and only a fraction is reused or recycled. That means, on average, you and I each throw away 54 pounds of textiles per year. Out the door with those clothes go tens of thousands of gallons of water. But we're not throwing away as much as we're buying. So in addition to piling up perfectly good clothing in landfills, we're also stockpiling our wardrobes. Most American women, for example, have more pairs of jeans than there are days of the week! And blue jeans are far from water lean.

Every new pair of jeans costs nearly 3,000 gallons of water to make. Given that 450 million pairs are sold annually in the United States, that comes to nearly 1.4 trillion gallons of water—the equivalent of half of California's entire yearly urban water demand. Why so much water? Well, there's the water to grow the cotton used to make the clothing, the water used to process the raw material in the textile mill, and then all the water needed to flush out the chemicals used in the field and in the factory.

Here's the breakdown: It takes about 2,247 gallons (a worldwide average) to grow enough cotton to make a pair of jeans, but it takes another 165 gallons to dilute the pesticides and fertilizers that go with cotton growing. The EPA stipulates that only low concentrations of pollutants can enter natural water systems, so additional water is added to runoff to make sure those standards are met.

Textile manufacturing facilities use different chemicals (and lots of water) to treat the raw cotton (97 gallons of water); bleach, dye, and print the fabric (95 gallons); and then finish it (about 36 gallons per pair). That's right, the effects of years of climbing, crouching, sitting, and dancing can now be simulated with a little caustic soda and a lot of water, getting you that perfect "worn" look and feel.

Sticking with jeans as an example, just one pair requires 232 gallons of water just to thin out the chemical stream. When Brooke Shields famously informed us that "nothing" comes between her and her jeans, she forgot to mention all that water! (Well, perhaps that would have been embarrassing.) Anyway, making denim out of organic cotton, hemp, or wool helps address some of these excessive water issues by producing less wastewater. However, organic jeans remain a small and expensive part of the clothing industry's main fare. Only 0.55 percent of worldwide cotton production is organic. And while cotton may be the most important natural clothing material (used for 40 percent of all textiles), it's not the only water culprit.

Sheep are dipped in chemicals to remove bugs before making

wool, and synthetic materials like polyester may require less water to "grow," but they generate volatile organic compounds (commonly called VOCs) and pollutants during the manufacturing process. All of this has to be washed (at least partially) out the fabric and diluted in the wastewater stream. With the high discarding and low recycling rates for clothing, much of the fabric eventually ends up in landfills, where pollutants from dyes can leach into the water supplies.

Buying new clothing has less to do with function and fit than it does with self-expression these days. And there is nothing wrong with that. In fact, clothing is a great way to express a whole lot more than just your fashion sense—you can use it to express your water consciousness, too.

Here's how to choose how much water you're carrying with that shirt on your back.

Jackets—7,956 gallons per hip-length jacket made from cowhide. Avoid leather if you can, and look for hemp or linen blends instead of cotton where possible. If you've got to go T-bird style but you don't need heavy-duty protection (e.g., if you're not a professional motorcycle racer), go for lightweight (often goatskin) or sheep- or lamb-based suede. Shorter is better, too.

Pants—2,866 gallons per pair of jeans (including bleaching, dying, printing, and finishing processes). A similar water footprint applies to corduroys and cargo pants, all of which are made of cotton. Wool and linen pants require less water than cotton trousers, as do hemp or hemp blends. When possible, do like they do in Bermuda: Go with shorts. (Pastels optional.)

Running sneakers—1,247 gallons per pair. Try to find ones manufactured in the United States to cut down on the costs associated with shipping. Doing that also helps ensure that the manufacturer is following fair labor practices, a notorious problem within the sneaker industry.

Shirts—975 gallons per cotton men's dress shirt. Instead of those made from 100 percent cotton, try to find blends containing hemp and/or linen, both of which have lower water footprints. By all means, avoid the fancy Egyptian cotton brands. Egypt uses 100 percent irrigation for its cotton, compared to only 52 percent in the United States.

Shoes—2,113 gallons per pair of leather shoes. Most leather shoes are made from cowhide, which has a much larger water footprint than many other animal materials—no matter what your shoe size is. Try to find canvas or hemp shoes, or sandals (less material) where possible. (Espadrilles used to be the height of fashion back in the day.) If you have to go with leather, follow the King's advice and go for suede— primarily made from sheep or lamb hide. It's soft, often cheaper, and has almost one-third the water footprint of tanned leather.

Socks—244 gallons per pair of socks. Cotton is one of the most water-intensive plants around, so when possible, try to find blends that incorporate hemp, flax, or bamboo fibers to reduce the water footprint. Unless you're a '70s basketball star, go for ankle socks. Less material means less waste. When it comes to socks, your own footprint is really the best: When the weather permits, lose the socks altogether.

Suits—3,900 gallons per cotton suit. You can reduce your water footprint by two-thirds by purchasing linen suits for warm-weather wearing. Choose wool suits for fall and winter. If you need something fancy, cashmere (made with goat hair) also consumes less water than cotton: only 1,528 gallons a suit.

Sweaters—594 gallons per wool sweater. Nothing beats wool for warmth and coziness. Its water footprint is less than cotton and petroleum-based synthetics such as fleece. Get 'em handmade, too. No longer your typical nightmare present from an estranged aunt, handmade sweaters are coming back in style with the revival of knitting groups. Support local artisans and keep warm, all while reducing the wasting of water.

T-shirts—569 gallons per cotton T-shirt. Choosing a lighter color and less print will reduce some of this heavy water footprint. Tank tops cut down on the total amount of material and, therefore, water use. The best choice is to look for tees made of hemp, which has one-quarter of cotton's water footprint but similar "breathability."

Underwear—86 gallons per ladies' regular cotton bikini style or 252 gallons for men's cotton boxers. Skimpier can be better for water when it comes to underwear, as less material means less water. Some of the sexier materials, such as silk, also take a fair amount of water to produce. So, ladies, go for the thong. If more coverage is your thing, try looking for cotton blends (a cotton-and-hemp blend is a good option) in boxers or briefs. If you're going to stick with cotton, look for those made in the USA.

BOTTOM LINE

Less is more. Shorts, short-sleeved shirts, and ankle socks may expose more skin, but it takes less water to make them. The more material, the more water used. Keep your clothing skimpy.

Material matters. Egyptian cotton means twice as much water is needed for irrigation as good ol' made-in-the-USA cotton, while wool is mostly a matter of shearing. Animal skins (like leather) are never cool as far as water goes: They require the most water.

Secondhand is suitable. Extending the life of a garment extends the water supply. Selecting vintage keeps more water in the ground and less on your back; there's no need for virgin materials to be manufactured.

8

furnishings

the H_2O of household items

We are creatures of comfort, forever seeking beds that bring better slumber, chairs that improve our posture, couches that support hours of veggin' out in front of the TV (complete with remote controls built into the armrests), and tables that allow us to host the perfect dinner party or poker night. And these are just the basics.

The average American home has 10,000 items in it. We need storage units and cabinets in just about every room to house all these things. But you would need an Olympic-size swimming pool to store all the water it takes to make what we so casually call "our stuff."

In fact, if you were to release all the water in all the items in our homes, at least 200,000 gallons would gush out like in some mad scene in a disaster movie. Americans expend a lot of water spending that $78.5 billion a year on furniture. Here's why: Most of the furniture we own (38 percent) is made of wood. Just 1 board foot of lumber takes about 5.4 gallons of water to make. Wood, of course, comes from trees, and last time I checked, it took water to grow trees. A full-grown pine tree, for example, needs thousands of gallons of water to mature to a size that's ready to be chopped down for timber. Most timber wood comes from trees more than a decade old—it takes that long to get big enough to be felled.

It's one thing if that timber is derived from trees in a sustainably farmed forest—one that is cared for and maintained so trees can grow anew for decades or centuries. The Forest Stewardship Council (FSC) certifies companies that adhere to responsible management processes. But most timber trees aren't so lucky. As much as 40 percent of the world's timber supply comes from illegal logging. These illegal loggers chop down trees without regard for the soil, the water tables, or the futures of the forests. They wreak havoc on the environment, ruining chances for new trees to grow and resulting in the loss of water that we cannot easily recapture—"green water."

Green water is precipitation that falls on trees, soil, and grassy areas but escapes our capture and direct use—it's stored in such forms as soil moisture, and trees can harvest it, but we cannot. Blue water is the water that ends up in rivers, lakes, aquifers, and other bodies of water that we can easily tap into.

The chair you are sitting in right now, for example, likely took more than 44,000 gallons of green water to make. If the trees it took to make that chair weren't sustainably harvested, then more trees won't grow in their place and that amount of green water in that region is lost.

Of course, wood is not the only material used for our furniture. Upholstery and metals comprise the next biggest components of furniture. These materials are no slouches when it comes to water consumption, either. Wool and leather used for upholstery come from sheep and cows that drink and eat, and metals come from mines that sift, process, and discharge vast amounts of water.

Buying secondhand furniture is great way to stem the drain on virgin materials. There's certainly plenty of "vintage" from which to choose. The US Environmental Protection Agency says there's an 8.8-million-ton pile of furniture at our landfills. So next time you hear some inane furniture-store ad claiming that its prices can't be beat (!), or that your mattress will be free (!!), or that you won't have to make payments until 2060 (!!!), remember that there are other prices to pay—environmental prices—for buying something new.

With that in mind, here are some ideas to help you relax and get comfortable with the water costs of your furniture purchases.

Beds—2,878.3 gallons per queen-size spring mattress. Sure, there are all sorts of new, "all natural" mattresses out there. But the water count skyrockets when you take into account the amount of natural fiber needed (more than 52,400 gallons for a cotton-and-wool mattress, about half that for hemp). Try a foam rubber mattress. It's not as bouncy, but it can be form fitting for comfort and uses less than 20 percent of the water needed for natural fiber beds.

Blankets—3,824 gallons per wool blanket. It takes about twice as much water to grow cotton as it does to raise a sheep, so wool works best for natural fibers. A similar-size acrylic or other synthetic material blanket takes much less, around 166 gallons. If you go the comforter route, you have to account for the down or fill inside, too—and that's tens of thousands of gallons of warm water to cover yourself with. (Geese gulp.)

Chairs—more than 11,000 gallons per leather chair, not including the wood or steel involved. Forget the leather recliner and go for something covered in rayon. It may be the least-water-intensive material for making a comfortable chair, and it biodegrades rather quickly. For more functional chairs, go with rattan or FSC-certified wood, if you must. They're harder on your butt, but softer on water.

Chest—91.5 gallons for a three-drawer chest. Consider this: To support every inch of a tree's diameter, the tree typically needs 10 gallons of water. Depending on the size, anywhere from 40 to 500 gallons of water are needed to feed a tree monthly. Get secondhand furniture. It saves all of those virgin resources. Think of it as cutting the cost in half every time stuff is passed along. Vintage reigns on water cost.

Couches—more than 35,600 gallons per leather couch (about 6 feet by 3 feet). Tanning leather is inordinately water intensive. And that doesn't include the water needed for the wood, metal, foam, and other fabrics used to make a couch. Try faux leather. You don't have to feed it.

Dinnerware—12.7 gallons of water per set of four ceramic dinner plates, salad plates, bowls, and mugs. Ceramic doesn't require much water to make, but try to avoid bone china. Bone china is made with bone ash, often from cattle. The water needed to raise cattle is bad enough in what's on top of your plate, never mind in it.

Glassware—17 cups for an 8-ounce glass, or about 2 gallons for a 16-glass set of 12- to 17-ounce tumblers. Just because fast-food restaurants serve in giant-size cups doesn't mean you need monster-size glasses at home, too. Refills are free at home anyway. You can cut down on your liquid intake by cutting down on the size of the glass in your hand. And that goes for beer mugs and wine glasses, too.

Pillows—458.5 gallons per pillow. That's for one filled with man-made, recycled fibers. A goose down pillow can take 14,026.2 gallons of water to make. If you go for the hard stuff, what you may sacrifice in soft comfort you'll gain in terms of the hard facts of water saved.

Pots and pans—56.8 gallons for a saucepan. Stainless steel is hard on water. The bigger the pot, the more water is needed in and "in" it. Use thinner steel pots and pans of the right size, which are usually smaller than the ones you think you need. You don't need to cook pasta in a lobster pot!

Rugs—9,531 gallons per 5- by 9-foot cotton-and-wool rug. Go with an artificial fiber such as rayon (not a synthetic that's made from petroleum). If you can afford it, get a silk rug. Other popular rug materials are jute, hemp, and flax, all of which need water in almost every stage of the manufacturing process. They need so much water, in fact, that it'd be like stepping on several thousand gallons' worth every time you crossed the room—though that's not exactly what is meant by "walking on water."

Sheets—6,663 gallons per set of queen-size 400-thread-count cotton sheets. Manufacturers tout higher thread counts as being more comfortable, but the lower the thread count, the less material used and the lighter the fabric. This means less water was used in the making. A queen-size 1,000-thread-count cotton set uses almost 9,000 gallons. The quality of the cotton fiber and the finish are more important than the number of threads woven in, anyway. You can rest easy knowing that.

Silverware—86 gallons per five-piece place set for four people plus five serving utensils. There's a difference between a dinner set and a place set: A dinner set has larger forks and knives. Go for the place set. If you need a bigger fork to take a bite, it's a good bet you're taking too big a mouthful! Also, silverware is usually marked "18/0," "18/8," or "18/10." The latter number of the pair is the percentage of nickel that's in the piece. The "18" is the amount of chromium

(18 percent) in it. Putting in the nickel takes more water, but because it only affects shine and rust prevention, you can ditch the nickel and save the water.

Tables—56.7 gallons per 5- by 3-foot dining room table. If you can, make sure to select FSC-certified wood. As much as 40 percent of the timber around the world is derived from illegally logged trees. FSC certification guarantees that the wood was harvested sustainably. That means the land wasn't ravaged and resources (including water) were accounted for. It isn't just what you eat that matters, it's also what you eat on.

BOTTOM LINE

Get back to basics in your home. Wooden chairs, wool blankets, and single-fiber rugs make for better-water decor. The more stuffing in your stuff, the more water you're using.

It's "sew" not right to go for higher thread counts. The more stitches, the more water. You won't be sacrificing comfort, though, just high prices and water costs.

Think lean in the kitchen—and not just for your diet. The thinner the pot or pan, the less water was needed to manufacture it. You should also note that while *steel* isn't spelled with an *a*, it can take more than its fair share of water when people use bigger items than they need to (such as using a pot when a saucepan will do!).

9

health and beauty

bail out your bathroom

Every year, Americans spend more on beauty products than they do on education. Just shy of $100 billion is spent annually on skin care ($24 billion), makeup ($18 billion), hair ($38 billion), and perfume ($15 billion) products. But while the newest age-defying moisturizer may erase fine lines as if by magic, more than just wrinkles is vanishing in the act—clean and abundant water is disappearing, too. If every American woman over the age of 15 owned just one lipstick, all those tubes would be holding more than 2.4 million gallons of water. And that's a very conservative estimate; most fashionable females have more than one shade in which they purse their pouty lips.

But the water wasted isn't just *inside* the tubes. Water goes into making the tubes, too. Because it's unlikely that we'll be seeing super-size bottles of mascara any time soon—most cosmetic products come in their own individual cases or plastic containers— there's a ton (actually, tons!) of extra packaging that burns fossil fuels, creates waste, and costs water. How much water? Well, about 3.4 million gallons to supply half the women between the ages of 20 and 64 in the United States with one small blush container. That isn't pretty.

And what happens when that blush brush starts scraping bottom? Discarded cosmetics are burned, buried, or flushed. Whichever way, chemicals can be released into the air, earth, or groundwater, depending on the incinerator, landfill, or water-treatment facilities. When that happens, that freshwater ceases to be fresh; it's polluted.

How do we know this? Personal care products flushed or rinsed down the drain have been linked to decreased fertility and skewed sexual development in everything from frogs to fish. If that's true, then you can bet the water isn't so healthy for us, either. Turns out that it ain't just the demand for his legs that has Kermit worried, it's an extreme makeover. And we may want to be concerned about that, too. Remember when it was discovered that pharmaceuticals flushed down the toilet end up in our drinking water supplies? Well, the same thing goes for cosmetics. Indeed, it's an ugly fact.

But we don't have to compromise our health or that of the planet to look good. Here's how to apply a water-conscious, animal-friendly, healthy beauty regimen that's pleasing to any eye.

Antiperspirants—220 gallons per 2.4-ounce stick that's 50 percent a waxy medium like castor oil. It's better for your water footprint, and possibly your health, to choose a roll-on with a water base. Antiperspirants act by clogging your pores to prevent sweating, and the chemicals used in these products may be associated with everything from Alzheimer's disease to cancer. Look for natural deodorants and cut back on water and your stink by swapping out some red meat for veggies. Fruits and vegetables create less body odor than burgers. Stick with roll-ons over sprays because they last longer and don't carry a risk of bursting into flames when applied.

Cosmetics—the amount of water involved varies by product. Many eye shadows, mascaras, and blushes contain waxes, oils, and pigments, so there's little water. However, many of the oils have high water footprints (such as castor oil, at 2,516 gallons per pound). Look for all-natural, mineral-based products and those with minimal packaging. Many major cosmetic companies will recycle old containers (and some even give you a freebie when you turn in your empties!). Recycling products is also a good idea. Leftover lipstick can be melted and remolded. If every woman reused just one lipstick, it could save more than 2.4 million gallons of water in the United States alone. Put 'em together and blow.

Moisturizers—3.6 fluid ounces of water per 6 fluid ounces, plus the amount in the packaging. To save on water and get the most out of your moisturizer, lube up right after showering, allowing the moisturizer to trap in the water on your damp skin. Moisturizers come in two types: water based and oil based. Oil-based products will keep your skin moist longer, but some oils have heavy water footprints. Go for coconut oils over cocoa butter, because that little *a* makes a big difference: 762 versus 6,808 gallons per pound! For sunburned and sensitive skin, look for aloe vera. This desert plant dishes out a lot of benefits without demanding much water.

Perfumes—1 ounce of water per 5 ounces of eau de toilette, 0.5 ounce per 5 ounces of cologne, and almost none for pure perfume. However, those figures belie the fact that huge quantities of water are used to make the hundreds of ingredients in a single fragrance. For example, you may have heard that fragrances made with essential oils are preferable to synthetic ones. But considering that it takes 300 pounds of rose petals to produce 1 ounce of essential rose oil, there's an entire bathtub full of water in that single drop of fragrance that you rub on your wrist. The ingredient hyraceum likely has the lowest water content of the perfume fragrances, but there's a catch: This smelly substance is actually the petrified urine of the rodentlike rock hyrax. Citrus scent, anyone?

Shampoos—about 17 ounces per 22-ounce bottle. Shampoos are 70 to 80 percent water, with some detergents, preservatives, fragrances, and moisturizers added. Try to avoid shampoos with jojoba, which takes 1,479.1 gallons of water per pound to produce. You can also save on water by shampooing only a few times a week—more than that strips your hair of its natural oils anyhow. For even greater water conservation, try one of the two-in-one shampoo-plus-conditioner products. And don't forget to turn off the tap while you lather up!

Soaps—180.4 gallons per 3.1-ounce bar. Look for castile-based (olive oil) soaps to reduce the high water footprint associated with soaps derived from beef tallow (who wants to wash with animal fat, anyway?). Tallow is listed as "sodium tallowate" on ingredient lists. Bar soaps will last longer than body washes and are often cheaper, by the way. Body washes are convenient, but they contain lots of water (and lots of chemicals to keep the goo gooey). They also have all that plastic packaging, which has its own water footprint, remember. Go for the bar and keep it in a covered dish so it doesn't dissolve away during showers and baths.

Toothpastes—about 2.1 ounces per 4.4-ounce tube. Toothpaste is about 50 percent water already, so there's no need to add more to get those pearly whites shining. The toothpaste contains all the ingredients necessary for creating a nice, foamy lather without wetting your brush first. You can save water, too, by only using a pea-size dab of toothpaste—your tube will last longer, as will the water table.

BOTTOM LINE

Go for the natural look. Many cosmetics use oils, one of them being castor, which burns through water. Other ingredients can also clog the water supply—and your pores. Healthy and natural beauty—the kind found, say, at birth—is easiest on the eye and requires the least water fetching to produce.

It's what's inside that matters, so avoid products with lots of packaging. It takes water to produce the plastic, glass, and paper used to contain beauty products. There's often more container than product in what you buy. If you use bar soap, however, no plastic container is required.

Don't flush your blush. It's an ugly truth: People dump unwanted cosmetics down the drain or flush 'em down the toilet. When these products enter the water supply, they can come back to make us all look bad by polluting the water, fish, and plants—and us.

10

school and office products

the (water) supply closet

All work and no play makes for a very watery day. Whether it's math homework, handouts, or a memo to the board, schools and offices chew through thousands of sheets of paper a day; an average office worker goes through 10,000 sheets a year! Those 20 reams leave a heavy water footprint behind—about 1,425 gallons.

Copying is what gets us in a lot of trouble. More than 4 million tons of copy paper are used each year, only one-third of which is made with recycled materials. Another third is from sawdust and wood chips, and the remaining third is from thousands of trees and more than 25 billion gallons of water. That's as bad (if not worse) than looking over someone's shoulder during a school exam.

The printing industry is one of the biggest industrial water consumers. More than 2 billion books are published each year in the United States, and textbooks (from grade school to college level) make up about 30 percent of the market.

Okay, I'm all for reading (especially this book), but still . . . A standard printing press may use more than 3,000 gallons per year, while some specialty forms of printing can exceed 25,000 gallons every month! Add to that the gallons needed to flush out all the chlorine used to bleach the paper (90 percent of all paper used in the United States is white), and it's a wonder the paper aisle isn't flooded.

But water leaves its mark not only in the paper but also on the paper. Whether ink cartridges are tossed in landfills or recycled, they create a toxic sludge that has to be treated with water. Enough to make you see red, isn't it? But stopping the presses won't plug the water problem, either.

Despite the technological revolution, both offices and schools still rely on good old-fashioned pencils. About 14 billion pencils are made every year, requiring about 6.5 billion gallons of water. That's enough to fill Chicago's Sears Tower (oh, sorry, the Willis Tower) more than 16 times!

But it isn't just the nerds with their books and pencils that rack up the water use in schools. Jocks are accountable, too. Every faucet and every football field use gallons. One school district in New Mexico uses more than 500 million gallons of water annually, much of which is to maintain its 300 acres of turf! Perhaps those football stars should start wearing Speedos to practice.

And graduates don't seem to have learned their lessons, either. Roughly 167 gallons of water are required for an average wooden desk. A small business with 12 cubicles will therefore spoon out more than 2,000 gallons of water just for the work surfaces! Water adds up in every stapler and paper clip, too—especially those that come packaged in plastic.

Doing a little homework or some research on overtime can go a long way toward cutting back on water (and environmental) waste, however. The good news is that you don't have to study or work too hard; I've provided you with many of the answers right here.

Books—42.8 gallons per book. Depending on the paper stock and the book's cover and size, a book—such as the one in your paws right now—can hold a lot of water between its covers. Look for those that aren't just good reads, but are also good for the world by virtue of being printed on recycled paper. Get bonus points if you buy them used.

Computers—10,556 to 42,267 gallons, depending on the type. The 2-gram microchip contributes 8.4 gallons alone. To reduce the water footprint of computers, look for reduced packaging and housing, and instead purchase a new one only when you really need it. I know, I know, the newest versions are always so flashy, but holding on to the one you have got and buying upgrades reduces demand for new materials. Look for products made by companies that are taking steps to green their entire supply chains by removing toxic chemicals, increasing recycling, and reducing packaging. Finally, in the world of water conservation, smaller is better: A laptop uses slightly more than half the amount of material that a conventional desktop computer does, requires less packaging, and consumes less energy, too. When it comes to your PC, think thin and slim.

Erasers—40.3 gallons per rubber eraser. The biggest mistake you can make in buying a rubber eraser is to buy a real rubber eraser! Most are made from synthetic materials anyway, and they require just 2.4 gallons each. Those made from soy-based materials also do the job just fine and don't make water disappear as fast.

Paper—3 cups per sheet. If ever there was a reason to recycle, it's because of paper. Paper is the number one thing you find in landfills. Using postconsumer recycled paper means you can save two-thirds of the water used to make new paper from trees. A typical 500-sheet ream of nonrecycled paper adds up to 93.7 gallons of water. Reduce. Reuse. Recycle!

Pencils—7.5 cups per pencil. Pencils aren't made of lead anymore, they're a mix of graphite and clay, plus wood, of course. The more clay, the harder the pencil. Likewise, the more clay, the more water used. Choose a lower number such as #2; they're medium soft.

Pens—1 gallon per pen. A disposable pen is a waste. Just because the ink has dried doesn't mean that all the liquid is gone from a pen; it takes more water to make the plastic. Get a reusable pen and replace the cartridge.

Printers—9,510.2 gallons for the average color printer. China and India are where most printers are manufactured. It happens that the cost of water used in manufacturing is four times less in these countries than in the USA—something we should copy.

Tape—6 cups per 12.5-yard roll of clear tape. Clear tape (the best in terms of water use) is made from cellulose that has been formed with polymers and wood pulp. Stick with that.

BOTTOM LINE

Reduce your use. We likely use the most paper at work or at school. Using e-mail instead of faxing or mailing letters and e-books instead of paper ones will stave off the need for cutting new paths through the forest. Trees need water to grow.

Reuse the products you have. It takes water to make new things such as pens. You don't have to dispose of a whole pen when a replacement cartridge is all you need. And this year's computer model, I'll bet, is just about the same as last year's, so how 'bout upgrading your computer's memory rather than the whole CPU?

Recycle the raft of paper (and other products) you collect at school and the office. Paper is the number one thing in landfills. In other words, it's what we toss the most of. Try giving it a second life on the front end by purchasing recycled paper products and on the back by using the recycle bin. More than 25 billion gallons of water per year are used for making virgin copy paper alone.

11

luxury

bling bilge

The rich are indeed different: They use a hell of a lot more water.

For a growing number of the super-rich, private jets, fancy sports cars, and five-star restaurants are but a drop in the bucket. Unfortunately, that drop can represent a tidal wave of water.

Take diamonds, for example. Mining and carving these gems requires thousands of gallons of water. The diamonds may be forever, but the water in rivers that are drained or polluted by mining to get them can be difficult to recapture. What's more, many diamonds are found in already-water-scarce regions of Africa. Each year, it takes 2.7 billion gallons of water just to pop the question in the United States alone. Every "Will you marry me?" comes with a 1,380-gallon price tag!

It isn't just the bling on the body that can get us into trouble. Classic one-upmanship in the yachting world has led to boats so big they now need to be assembled in industrial facilities designed for building military ships. Today, there are more than 2,000 yachts exceeding 100 feet in length. Construction of such floating mansions requires so many millions of gallons of water that I couldn't calculate it all. To be sure, these megayachts could float on all the water it takes to make them.

Land toys can be equally harmful. A Lamborghini Murcielago gets a paltry 9 miles to the gallon, almost 25 percent less than a Hummer. Its mpg is low, and its WPC (water per car) is high. For an average car: 39,000 gallons. Add to that the extra fuel needed because of the poor mpg, and water is fast and furiously gone.

And then there's the rich water above. No, not clouds! Private jets wreak so much water havoc in the sky that it's difficult to know where to begin. Construction and fuel mean millions of gallons lost. But jets also affect H_2O in a different way: They move water particles around. More than 5 gallons of water vapor are released with every mile flown. That means that molecules that might have fallen on land to help us water our crops instead might be lost over the sea.

Kinda makes you want to have a drink, right? So what about the water footprint of that wine cellar? A "downsized" 2,000-bottle collection represents more than 300,000 gallons. That's a lot of water to be hiding in the basement. But, sipping your favorite champagne doesn't have to mean draining the aquifer. Here are a few essentials to swill that will help you reduce the effect of those extravagant tastes you have.

Cars—39,000 gallons. And that's just for the steel. Cars, like most industrial products, use a lot of water per dollar spent. For example, the global average water footprint of industrial products is 80 liters per dollar of purchase price. In the United States, industrial products take nearly 100 liters per dollar; in Germany and the Netherlands, it's about 50 liters per dollar; in Japan, Australia, and Canada, it's 10 to 15 liters per dollar; and in the world's largest developing nations, China and India, the average water footprint of industrial products is 20 to 25 liters per dollar. The difference is explained by the portion of water resources used for industry compared with the value of a country's industrial sector. The United States obviously has a highly valuable industrial sector that creates high water demand as well—enough to drive us toward drought.

Jets—about 1 billion to 2 billion gallons, depending on their customization and size (did you know private jets come in tall, grande, and venti?). Opting out of leather seats will help cut the water footprint, as will removing all those bottles of water and installing a small water tank. But face it, it's tough to make a dent in more than a billion gallons of water (see "Cars" for a discussion on water use in making industrial products). If you have to fly private, consider fractional ownership, where you and other companies or individuals share the same fleet. At least then the initial water footprint of the jet is spread amongst all you high-flying yahoos.

Jewelry—2,010 gallons for a pair of 1-carat diamond studs. The message with jewelry is simple: Save water by choosing recycled materials. Mining for gold, silver, and gemstones destroys the environment by, among other insults, polluting freshwater sources with toxic waste products like mercury. Processing the raw materials also requires water. Today, trendsetters can choose from recycled stainless steel instead of silver or gold, as well as antique stones that came out of the ground and went into the machine shop decades ago. After all, one of the reasons why these rocks are such gems is that they do not degrade over time. Ain't that precious.

Televisions—3,900 to 65,500 gallons per TV, depending on its size, make, and model. The flatter and smaller, the better, because less material = less waste, including water.

LCDs tend to be lighter and more energy efficient than plasmas. Although shiny black models are trendiest, look for "green" TVs made by companies that consider their environmental impact from cradle to grave. The best water (and other resource) conservation comes from simply sticking with what you have: If your TV doesn't have rabbit ears or a manual dial for the channels, there's no need for an upgrade yet. Bigger isn't always better; instead, it's what you watch that will really keep the neighbors talking.

Watches—about 1,800 gallons or more for a quartz-driven watch. For the band, avoid synthetic rubber and leather, and also skip the cheap plastic (which won't last as long). It's best to look for models made with recycled stainless steel. Time's running out on our landfills, so invest in solar or mechanically powered watches to reduce the waste associated with making and disposing of batteries. Ticktock.

Yachts—millions of gallons. Building a low-end, 120-foot yacht with a price tag of $13 million requires more than 275 million gallons of water, based on the amount of money to the amount of water ratio (those 80 liters of water, on average, around the world per dollar cost of an item). You could probably set sail on all that water! Yachts are big and require lots of materials, from fiberglass to steel to wood. Fiberglass has a much lower water footprint than steel, so for small boats, go for those over a metal rowboat. If you're going sailing, do it the old-fashioned way: Hemp ropes and sails used to outfit every ship sailing the western seas prior to 1850. In general, Canadian industrial products have about one-tenth the water footprint as American-made. When it comes to these big items, you can save water by buying your boat from a Canadian boatbuilder and sailing it home. Bon voyage!

BOTTOM LINE

Downsize. The bigger and more expensive the product, the
more water it likely requires. Industrial products made in the
USA take an average of 100 liters per dollar to produce.
Being economical extends to water costs as well.

Old bling, as in antique jewelry, is better than the new stuff.
Mining creates toxic wastewater regardless of whether it's for
gold or diamonds. Remember, these things are meant to last
forever, not just one relationship.

Boats comprise boatloads of water. Hitting the high seas
under sail is the way to go. Russian billionaires take note:
Motor yachts require so much water to build that it couldn't
be figured out.

12

pets

the irony of fire hydrants

There are more than 77.5 million dogs, 93.6 million cats, 15 million birds, and 13.3 million horses in the United States, not to mention other exotic pets like chimps, pigs, and snakes. And we spend a lot of time and money on all of them—$45.4 billion annually; that's $730 a year for each canine alone.

I always say that if I were to come back into this world as another living creature, I'd want to be my dog. He gets fed, walked, picked up after, bathed, and even massaged. When I go out of town, he stays at a pet resort. Still, even though he is a "water dog," I never particularly thought about his effect on water—his water pawprint, if you will—beyond the bowl I keep fresh for him to drink from. But if you add up all the dogs in the country, they lap up more than 18 million gallons of water per day. Add to those the bowlfuls for the feline population, the beakfuls for the birds, and the bottle- and troughfuls for the rest and, well, our animal companions make a significant water mark on the world.

Speaking of marking, it's one of the ramifications of drinking and eating (let's call it fertilizer) that makes pets' biggest effect on water. More than one-third of all dog owners—38 percent—don't pick up after their pets. That leads to water contamination, rendering water sources unhealthy and unusable. In fact, pet waste is estimated to cause between 20 and 30 percent of the water pollution in the United States. It happens more often than you might think.

At some California beaches, dog feces picked up by the tide contribute about 10 percent of the *Escherichia coli* bacteria in the water—enough to make swimming unsafe. Near our nation's capital, in the watershed of the famed Four Mile Run Park in Fairfax and Arlington, Virginia, dogs leave behind more than 5,000 pounds of "waste" every day, adding to the contamination of Washington, DC, streams. Nationwide, dog pooh piles even higher. An estimated 3.6 billion pounds—enough to pile 800 football fields a foot high—are deposited every year and end up either in landfills or in areas where the bacteria eventually make their way into groundwater, drains, or streams. But dogs aren't the only culprits: Cat litter gets tossed as well, or flushed down the toilet—which is really harmful for the water supply because it clogs toilets and wastes water. Needless to say, all pets create waste. It's how we handle it—or not—that causes problems.

Pet products also pounce on water supplies. We spend almost $30 billion on pet products and supplies every year, never mind the $15 billion or so we spend on pet services. Take pet food, which is largely made of grains, meats, and rice—all of which take oodles of water to grow or raise. And then there are all the toys, the toys! The stuffed pheasants, mice, beavers, weird octopuses, dodos, etc., all require energy, water, and artificial (let's hope) materials to be made.

A new trend that has a dramatic impact on water is the popularity of upscale pet products and services. According to the American Pet Products Association (APPA), "High-end items to spoil companion animals are must-haves for pet owners that spare no expense to please their furry, feathered and finned best friends. Items include faux mink coats for cold weather outings, feathered French day beds for afternoon naps, designer bird cages, botanical fragrances and to top it all off, a rhinestone tiara!" Seriously.

Think about all this in the context of water: "Pet-owners take grooming one step beyond a haircut, a quick bath and a nail trim. Mouthwash and an electric toothbrush for canines are routine steps in a beauty session for some pooches. Birds receive daily pedicures with special cage perches, while others enjoy manicures complete with nail polish. Pet-owning homes stay cleaner with automatic, self-flushing litter boxes, cleaning cloths for muddy paws that mimic traditional baby wipes, and scented gel air fresheners to keep rooms free of pet odors." It gets better.

The APPA says, "As pet owners meditate in yoga class, cats relieve stress by frolicking in a toy gym or relaxing in a feline spa before enjoying herbal catnip packaged in a tea bag. Dogs sip fresh water from flowing fountains after a soothing rub with a doggie massager." As mentioned, in another life . . .

Meanwhile, here's some pet-friendly, water-saving advice.

Pet beds—1,654 gallons or more for a 12-pound, medium-size bed. While many pet beds are filled with polyester or other artificial materials that consume lots of water in the making, smarter choices that use recycled plastic bottles are available. Shredded and stuffed into a canvas cover, these beds save up to 92 gallons of water each by keeping plastic bottles from going to waste—but never degrading, let's remember—in landfills.

Pet bowls—2.5 gallons for a medium-size ceramic bowl. I bet you don't like eating off tin or plastic, so why should your pet? Tin is noisy and plastic is too lightweight. Besides, processing those materials takes more water than ceramic does. Keep the water for inside the bowl and not the bowl itself. Also, you can skip those that are colored with dyes and painted with

cute little sayings. Your pet can't read. If you think he or she
can, call David Letterman. Stupid Pet Tricks awaits.

**Pet collars—248.6 or more gallons for a medium-size
leather dog collar.** You can waste hundreds of gallons—and
at some stores hundreds of dollars—on a leather collar when
a hemp one will do just fine by the wallet and the water
supply. Even better, let your pet run naked. Unless you're in
an area that requires a license or your pet likes to play "the
great escape" game, you don't need a collar at all.

**Pet foods—up to 460 gallons per pound of dry vegetarian food
to 1,580 gallons per pound for meaty canned food.** Dry
food (kibble) contains only 10 to 12 percent water. And if you go
with all vegetables, your pet will have an even smaller water
pawprint. Dry food lives up to its name by requiring three times
less water for processing than canned "stews" or "sauces" and
the like. Government regulations actually require that canned
pet food be no more than 78 percent moisture. Try "raw" food
that beckons your pet back to his or her species' "wild" days
for less processing and less water in the mix. Meow.

Pet leashes—17.6 gallons for a 6-foot nylon leash. The
simplest way to save water: Keep your pet on a tight—as in
short—leash. The less material, the less water, and the less
liable your pet will be to wander into trouble. Your neighbor's
new flower bed will thank you.

**Pet toys—250.3 gallons for an 18-inch length of hemp rope
2 inches thick.** Cats love to chase strings. Dogs like to tug on
ropes. Why not entertain with simple and natural things rather
than some windup toy animal that isn't fooling anyone but still
consumes thousands of gallons of water in the making? Worse
are those that are battery operated. There are lots of eco pet
toys available that use wood or organic fibers that are better
for your pet's health and better for the planet, too.

Pet treats—395.2 gallons for a 12-inch bully stick. A bully stick is actually a bull's pizzle, which can be several feet long. Seeing as you have to take down the whole bull to get it, or otherwise risk a painful adventure (and probably a quite surprising one for the bull), a lot of water has to go into the raising of the steer. It's safer to go with a small, dry veggie treat. How 'bout the relatively harmless potato chew? Spuds don't require that much water to grow, nor do they charge at you.

Poop bags—5 ounces per plastic bag. Although they are environmental scourges because they don't biodegrade, plastic bags require the least amount of water compared to all the biodegradable ones on the market to pick up after your pooch. Biodegradable bags are largely made of corn and can require more than 35 ounces of water per bag to manufacture. Holy crap!

BOTTOM LINE

Pick up after your pet. The courtesy isn't just so people won't step in it, it also prevents waste from entering the water stream and damaging supplies. As much as a third of the water pollution in the United States can be traced back to bacteria from pet waste.

Dry pet food actually keeps things wet. It requires less water to make. The water bowl shouldn't be confused with the food bowl, and both, by the way, should be made of ceramic.

Stop with the wacko pet toys. Give your dog a bone or your cat a ball of twine to play with. Artificially made products (and collars and leashes) should be met with a firm "No!"

13

building materials and appliances

it takes a lake to lay a foundation

The main purpose of shelter is to provide protection from harm, thievery, and the natural elements—especially rogue water. The last thing you want is a leaky roof soaking your carpet, rotting your walls, buckling your floors, and short-circuiting your electronics. Yet with all that we do to keep things dry, we almost never think about the water involved in the production and manufacturing of our dwellings themselves.

Indeed, every aspect of home construction involves water. From the cement that secures the foundation to the paint that coats the walls, you're living your life in an ocean of embedded water. It may be hard to believe, but the water used to produce your house is as much as your family will use in more than 15 years of living there. And how many people actually stay in the same home for that long anymore? The average person moves every 5 to 7 years.

We may pass along the water we used building our homes to the next occupants, but there is no reason to drown them. The key to water savings in building projects is knowledge—just like it is in all other areas discussed in this book. Knowing how much water is in different types of materials creates opportunities to make better choices. And when it comes to building, we have the opportunity to choose wisely literally from the ground up. Instead of building a watery abode, we can build a nice, safe, dry place to call home.

Given how many homes are built each year, those choices can add up to a lot of water savings.

The nearly 310 million people living in the United States reside in more than 120 million homes, and 1.5 million new homes are constructed annually. The average 1,700-square-foot home uses 9,726 board feet of lumber, 302 pounds of nails, 110 tons of concrete, 6,484 square feet of drywall, and 55 gallons of paint. The grand total of embedded water in your home—1.6 million gallons—could fill 2.5 Olympic-size swimming pools. Do the math and the grand total for all the new homes in the United States every year is 240 billion gallons. That's nearly enough to cover the entire state of Rhode Island in water 1 foot deep—every year!

If we were to squeegee all that water from all the houses ever built, the world would look more like *Waterworld* than *Sahara*.

Now, of the components that comprise a house, of course there are some nonnegotiables. You need to have walls and a roof, for instance. You need flooring, unless you're keen on walking around on bare earth. You'll probably sleep more soundly with some windows and improve your thermal comfort if you pack your attic with insulation.

All this aside, there are choices you can make to reduce your virtual water consumption by being picky about the materials you use when you're turning your house into a home. And, of course, you can build with efficiency in mind by selecting plumbing systems, landscape designs, appliances, and fixtures that are made to conserve water once the house is occupied.

For more information on conserving water in the home, check out the tips in Chapter 1. For now, strap on your tool belt (if it's made of leather, it contains about 11,000 gallons of embedded water, by the way), grab your hammer (another 15.5 gallons), and dive into the makings of your home.

Cabinetry—400 gallons for about 75 board feet of solid wood, the average amount required for a kitchen. Up to 1,590 gallons of water are used to create the same quantity of medium-density fiberboard (MDF), a wood substitute made by mixing wood waste and sawdust with synthetic resins (often containing formaldehyde) that is commonly used in cabinetry. MDF offers the appearance of wood for a fraction of the dollars, but with a high water cost and harmful pollutants. If you go for pure wood, look for cabinets that are Forest Stewardship Council (FSC) certified. If you choose MDF, pick one that has a low level of volatile organic compounds or is made without formaldehyde glues.

Carpeting—14,750 gallons for 1,000 square feet of synthetic carpeting, most of which is made of polyester or nylon. By choosing an equally chic carpet made of recycled plastic bottles, you'll cut the embedded water content by 86 percent—and keep more than 4,400 two-liter bottles out of the landfill.

Clothes dryers—16,909 gallons for the average clothes dryer made of steel, plastic, aluminum, and copper. Your best bet is to dry your clothes the old-fashioned way: in the sunshine. Buy a clothesline or rack—or just hang your wet clothes on hangers in bedroom doorways. They'll take a little longer to dry, but they'll last longer overall. Dryers can shrink clothes, and their intense heat is hard on materials. Lighten up, and by that I mean be breezy.

Granite countertops—3,920 gallons for 60 square feet of granite slab or tile (about 1 inch thick). Granite quarrying consumes 310 gallons of water per ton, while the processing of it uses 9,500 gallons. One of the big issues with some types of granite is radon emissions (which is a nice way of saying that your countertop is radioactive). You may want to bring your Geiger counter to the store.

Gutters—1,640 gallons for 120 linear feet of aluminum. Gutters installed along the outer rim of a structure collect rainwater as it runs off the roof and funnel the water away from the house. Originally constructed of wood and later made of copper or steel, most rain gutters are now made of aluminum—which is a good thing. Copper gutters cost more money and an extra 3,000 gallons of water. A penny saved in this case means water saved as well.

Hardwood flooring—7,200 gallons for 1,000 square feet on average (but with more than 50 different tree species to choose from, the range is quite large). From an environmental standpoint, it's best to avoid tropical woods, as they are typically associated with the clear-cutting of rain forests. It's best to buy wood certified by the FSC, which ensures that trees are managed and harvested in an ecologically sound manner. Almost half of all wood products come from illegal timber operations, and these operators are stealing water as well as wood.

Insulation—50 gallons for 2,500 square feet of shredded-newspaper insulation. In addition to the advantages of using recycled materials, shredded newspaper is preferable to fiberglass insulation because the latter requires four times more water. Sure, it's probably the least sexy part of building or remodeling, but insulation is hugely important for maintaining a comfortable and energy-efficient living space. Get your newspapers while newspapers still exist!

Linoleum flooring—3,865 gallons for 1,000 square feet. Made of all-natural materials like linseed oil, pine rosin, sawdust, and cork dust, linoleum is the preferred alternative to vinyl flooring, which is manufactured from chlorine and toxic plasticizers known as phthalates. Producing enough vinyl flooring to cover 1,000 square feet consumes more than 13,330 gallons of water. So go lino!

Ovens—13,738 gallons for an average range oven, the majority of which is consumed in extracting and processing the metals (iron and copper) that make up the unit. Try something smaller . . . and maybe faster. And let's call that thing a microwave.

Piping—6,930 gallons for 280 feet of copper piping (roughly the amount in the average 1,700-square-foot home). Copper piping is extremely resistant to corrosion, which is an important characteristic if you're trying to conserve water and minimize leaks. Since the 1950s, more than 7 million miles of copper tubing have been installed in plumbing systems across the country. With leaks accounting for an average of 11,000 gallons of water wasted in a home per year, a whole lot more copper (or at least faucet tightening) is needed.

Refrigerators—25,363.2 gallons for a standard side-by-side refrigerator. Refrigerators are primarily composed of steel, but also contain plastic, aluminum, copper, glass, and various chemical refrigerants. Ditch those frozen TV dinners and buy fresh food more frequently and you won't need such a big fridge!

Roofing—6,800 gallons for 2,000 square feet of cement tiles and 15,760 gallons for an equal area of clay tiles.

Although tile roofs are significantly more expensive than their composite-shingle counterparts, they look much classier and last at least twice as long. Composite roofs are made of asphalt, which is a sludgy by-product of crude oil refining, and have a water footprint of more than 39,844 gallons for 2,000 square feet of roofing. Best to keep the water off your head with tiles.

Stone tile flooring—58,800 gallons for 1,000 square feet of limestone or travertine tiles. Quarrying limestone consumes 600 gallons of water per ton, while processing it uses more than 30 times more (19,000 gallons per ton). Limestone is a naturally occurring sedimentary rock composed of minerals, fine sediments, and the shells of marine creatures. When limestone forms on land, it's called travertine.

Tile countertops—120 gallons for 60 square feet. Although ceramic tile is an energy-intensive building material because of the firing process, its manufacturing requires relatively little water. The best option is tile made from recycled ceramics or, better yet, tile made from recycled glass.

Washing machines—10,565 gallons for the average clothes washer made primarily of steel, along with some aluminum, copper, porcelain, and plastic, depending on the manufacturer. Compared to the water it takes to make them, washing machines consume far more during their lifetime of use. Over 10 years, a top-loader will use 157,000 gallons of water, while a front-loader will consume roughly 57,625 gallons. The best way to conserve water with your washer is to connect the drain hose to a drip-irrigation line to water outdoor landscaping. Just be sure to use plant-based detergents.

Windows—691.4 gallons for wooden window frames, 11,700 gallons for vinyl frames, and 19,190 gallons for aluminum frames. The primary smelting process used to make aluminum consumes huge amounts of energy and water and releases dangerous pollutants. Vinyl (or PVC, polyvinyl chloride) may also release toxins throughout its lifecycle. In addition to the embedded water savings, wood is the only material that's both renewable and recyclable (aluminum is recyclable, but not renewable). But wood is the preferred choice only if it's derived from a sustainably managed forest. Look for the FSC certification stamp; it's a good way to see your way through to water savings.

BOTTOM LINE

Make things with things made from recyclables. "Green" design innovators are coming up with building materials produced from all sorts of stuff, from plastic bottles to cardboard boxes and old newspapers. Materials made from recycled plastic is an especially good option because plastic doesn't biodegrade and the water savings can add up to 86 percent versus other building materials.

Avoid tropical woods and look for FSC-certified lumber. Forty percent of all the timber in the world comes from illegally forested areas, which means that water rights and sustainable management programs aren't being adhered to. That means water and lots of paradises are being lost to reckless operators.

Turns out that that 1970's staple of flooring, linoleum, is made from all-natural materials and needs a fraction of the water that vinyl does to be manufactured. Yeah, baby!

section
3

the sum of your water existence

adding it up

14

the water footprint calculator

liquid math

By now it should be clear that even if you're a Pisces, you don't have to leave a big water footprint. Taking a few simple steps can go a long way toward drying out those wet shoes of yours. Now it's time to put into practice all the tips you've just read and calculate the size of that water mark you leave behind.

Below is a short list of questions that lead you into some very basic calculations. Don't be alarmed. For this liquid math, all you need is a simple calculator, or, if you're a fan of arithmetic, a pencil and paper! Fill in the blanks, punch the numbers, and see how your choices add up. The goal is to get a rough idea of how much water you use so you can compare yourself to others and see how your choices can help you save water over time.

Taking the total from each question, you'll fill in the tables along the way—and the big one at the end to see the overall size of your water footprint. You should also be able to identify the areas where you can collect the biggest water savings. For a more detailed calculator, check out the online version at www. thegreenbluebook.com.

WATER CALCULATOR
FOOD AND DRINK
BREAKFAST

1. **Water footprint of your cereal bowl (including 1 cup of milk):**
   Allow 50 gallons per bowl of cornflakes; 115 gallons per cup of granola; 65 gallons per bowl of oatmeal; 95 gallons per bowl of rice-based cereal

   1a. Multiply (____) bowls of cereal per week × (____) gallons per type of cereal × (____) number of weeks per year = (____)

2. **Water footprint of bacon and eggs:**
   Allow 22 gallons per egg; 42 gallons per strip of bacon

   2a. {22 gallons × (____) number of eggs}
   + {42 gallons × (____) number of bacon strips}
   = (____)

   2b. Multiply total (____) × (____) number of times per week you have this meal × (____) number of weeks per year = (____)

3. **Water footprint of a cup of fruit and yogurt:**
   Allow 37 gallons per 6 ounces of yogurt; 4 gallons per cup of strawberries; 14 gallons per cup of blueberries; 18 gallons per banana

   3a. For each item, multiply (____) gallons × (____) amount per day × (____) number of days per week × (____) number of weeks per year = (____)

   3b. Add the results for all items = (____)

4. **Water footprint of toast, tea, coffee, and juice:**
   Allow 13 gallons for 2 slices of wheat bread; 17 gallons per 8 ounces of orange juice; 38 gallons per cup of coffee; 6 gallons per cup of black tea

4a. For each item, multiply (____) gallons × (____) amount per day × (____) number of days per week × (____) number of weeks per year = (____)

4b. Add the results for all items = (____)

Breakfast	
QUESTION	WATER FOOTPRINT
1a	
2b	
3b	
4b	
YEARLY TOTAL	

LUNCH

5. Water footprint of a typical sandwich:

Allow 54 gallons per 3 ounces of turkey; 88 gallons per 3 ounces of chicken; 296 gallons per 3 ounces of beef; 1.5 gallons per ½ cup of lettuce; 0.5 gallon per ½ tomato; 14 gallons per ⅓ avocado; 17 gallons per slice of cheese; 13 gallons per 2 slices of wheat bread

5a. Add gallons for all ingredients of your sandwich = (____)

5b. Multiply 5a (____) × (____) number of days per week you eat a sandwich × (____) number of weeks per year = (____)

6. Water footprint of a typical salad:

Allow 3 gallons for 1 cup of lettuce; 21 gallons for ½ avocado; 1 gallon per tomato; 6 gallons per ¼ cucumber; (____) gallons per other ingredient (see Chapter 6 for vegetables)

6a. Add up the gallons for all the ingredients in your salad
= (____)

6b. Multiply 6a (____) × (____) number of days per week
you eat a salad × (____) number of weeks per year
= (____)

7. Water footprint of soda:
Allow 39 gallons per 20-ounce plastic bottle

7a. Multiply 39 gallons × (____) number of bottles per day
× (____) number of days per week × (____) number of
weeks per year = (____)

Lunch	
QUESTION	WATER FOOTPRINT
5b	
6b	
7a	
YEARLY TOTAL	

DINNER

8. Water footprint of pasta:
Allow 38 gallons per serving of pasta

8a. Multiply 38 gallons × (____) number of bowls of pasta
per week × (____) number of weeks per year = (____)

9. Water footprint of rice:
Allow 50 gallons per ½ cup uncooked rice (about 1½ cups
cooked)

9a. Multiply 50 gallons × (____) number of times per
week you eat rice × (____) number of weeks per year
= (____)

10. **Water footprint of red meat:**

Allow 395 gallons per 4-ounce hamburger; 593 gallons per 6-ounce steak

10a. Multiply 395 × (____) number of burgers you eat per week × (____) number of weeks per year = (____)

10b. Multiply 593 × (____) number of steaks you eat per week × (____) number of weeks per year = (____)

10c. Add: 10a (____) + 10b (____) = (____)

11. **Water footprint of chicken:**

Allow 176 gallons per 6 ounces of chicken

11a. Multiply 176 × (____) number of days per week you eat chicken × (____) number of weeks per year = (____)

12. **Water footprint of turkey:**

Allow 107 gallons per 6 ounces of turkey

12a. Multiply 107 × (____) number of days per week you eat turkey × (____) number of weeks per year = (____)

13. **Water footprint of pork:**

Allow 230 gallons per 6 ounces of pork

13a. Multiply 230 × (____) number of days per week you eat pork (excluding bacon for breakfast) × (____) number of weeks per year you eat pork = (____)

14. **Water footprint of vegetables (1 serving = 1 side salad or palm-size portion of vegetables):**

Allow 3 gallons per 4 ounces of spinach; 2 gallons per 4 ounces of carrots; 10 gallons per cup of peas; 7 gallons per 4 ounces of broccoli; 6 gallons per serving of other vegetables

14a. Multiply (____) gallons per vegetable × (____) number of servings per week you consume of each vegetable

14b. Add the results for each vegetable in 14a = (___)

14c. Multiply 14b × (___) number of weeks per year = (___)

Dinner	
QUESTION	WATER FOOTPRINT
8a	
9a	
10c	
11a	
12a	
13a	
14c	
YEARLY TOTAL	

SNACKS AND SUNDRIES

15. **Water footprint of fruits (1 serving = 1 apple, 1 cup of fruit salad, or a 4-ounce smoothie):**
Allow 19 gallons per apple; 13 gallons per orange; 18 gallons per banana; 11 gallons per peach or nectarine; 4 gallons per cup of strawberries; 14 gallons per cup of blueberries; 15 gallons per melon; 15 gallons per serving of other fruits

    15a. Multiply (___) gallons per fruit serving × (___) number of servings per week you consume of each fruit

    15b. Add the results for each fruit in 15a = (___)

    15c. Multiply 15b × (___) number of weeks per year = (___)

16. **Water footprint of potato chips:**
Allow 50 gallons per 200-gram bag

16a. Multiply 50 gallons × (___) number of bags per week × (___) number of weeks per year = (___)

17. Water footprint of wine:

Allow 30 gallons per 4-ounce glass of wine

17a. Multiply 30 × (___) number of glasses per week × (___) number of weeks per year = (___)

18. Water footprint of beer:

Allow 20 gallons per 8-ounce glass

18a. Multiply 20 × (___) number of glasses per week × (___) number of weeks per year = (___)

19. Water footprint of bottled water:

Allow 4 liters per 1-liter bottle

19a. Multiply 4 × (___) number of bottles per day × (___) number of days per week × (___) number of weeks per year = (___)

Snacks	
QUESTION	**WATER FOOTPRINT**
15c	
16a	
17a	
18a	
19a	
YEARLY TOTAL	

Food and Drink				
BREAKFAST	**LUNCH**	**DINNER**	**SNACKS**	**YEARLY TOTAL**

HOME AND GARDEN

20. **Water footprint of bathing:**
    Allow 2.5 gallons per minute of showering; 60 gallons per average bath

    20a. Multiply 2.5 × (____) number of minutes per shower × (____) number of showers per week × (____) number of weeks per year = (____). Divide by 2 if you have a water-saving fixture on your showerhead.

    20b. Multiply 60 × (____) number of baths per week × (____) number of weeks per year = (____)

    20c. Add: 20a (____)+ 20b (____) = (____)

21. **Water footprint of brushing your teeth or shaving:**
    Allow 3 gallons per minute of a regular faucet running; 1.5 gallons per minute of a water-saving faucet running; 5 tablespoons per toothbrushing session with the tap off; 3 cups for shaving if you use a container to rinse the blade

    21a. Toothbrushing with tap running: Multiply (____) gallons per minute × (____) number of minutes × (____) number of times per day × 365 days per year (because you *should* brush every day!) = (____)

    21b. Shaving with tap running: Multiply (____) gallons per minute × (____) number of minutes × (____) number of days per week × (____) number of weeks per year = (____)

    21c. Add: 21a (____) + 21b (____) = (____)

22. **Water footprint of using the toilet:**
    Allow 5 gallons per flush; 3.5 gallons per flush with a water-saving device installed

22a. Multiply (____) gallons × (____) number of flushes per day × 365 days per year = (____) .

23. Water footprint of laundry:

Allow 27 gallons per wash in a front-loading machine; 40 gallons per wash in a top-loading machine; 20 gallons per wash in a water-efficient machine

23a. Multiply (____) gallons per wash × (____) number of washes per week × (____) number of weeks per year = (____)

24. Water footprint of dishwashing:

Allow 11 gallons per dishwasher load; 22 gallons if washing by hand

24a. Multiply (____) gallons × (____) number of loads per day × (____) number of days per week × (____) number of weeks per year = (____)

25. Water footprint of houseplants:

Notice how much water you give each plant each time you water it.

25a. Multiply (____) cups of water per plant × (____) number of plants × (____) number of times watered per week × (____) number of weeks per year = (____)

26. Water footprint of garden:

Allow 67,000 gallons per $\frac{1}{5}$-acre per week (in summer)

26a. Multiply 67,000 gallons × (____) number of times per week you water × (____) number of weeks per year you water = (____). If your lawn is much bigger than $\frac{1}{5}$ acre, increase the total accordingly.

26b. If you xeriscape or use drip irrigation, multiply 26a by 0.5 = (____). (If you do not xeriscape or use drip irrigation, 26b = 0.)

26c. Subtract: 26a (____) − 26b (____) = (____)

27. Water footprint of a pool:

Allow 19,000 gallons for an average-size pool; allow 250 gallons lost to evaporation per week

27a. Multiply 19,000 gallons × (____) number of times per year you empty your pool (use a fraction if you empty the pool every few years) = (____)

27b. Multiply 250 gallons × (____) number of weeks the pool is left *uncovered* per year

27c. Add: 27a (____) + 27b (____) = (____)

Home and Garden	
QUESTION	WATER FOOTPRINT
20c	
21c	
22a	
23a	
24a	
25a	
26c	
27c	
YEARLY TOTAL	

LIFESTYLE

28. Water footprint of clothing:

Allow 3,000 gallons per pair of jeans; 570 gallons per T-shirt

28a. Multiply 3,000 × (____) number of pairs of jeans you buy each year (use a fraction if you buy one pair every few years) = (____)

28b. Multiply 570 × (___) number of T-shirts you buy each year = (___)

28c. See clothes section in Chapter 7 to add in the water footprints of other items you buy every year.

28d. Add: 28a (___) + 28b (___) + 28c (___) = (___)

29. Water footprint of cars:

Allow 39,000 gallons per car

29a. Multiply 39,000 gallons × (___) the number of times a year you buy a new car (use a fraction if you buy a new car every few years) = (___)

30. Water footprint of paper:

Allow 19 gallons per pound of virgin (nonrecycled) paper; 6 gallons for 100 percent recycled paper

30a. Multiply (___) gallons × (___) number of pounds of paper you use per month (1 ream of average copy paper weighs about 5 pounds) × (___) number of months per year = (___)

31. Water footprint of plastic:

Allow 24 gallons per pound of plastic

31a. Multiply 24 × (___) number of pounds of plastic you recycle per week × (___) number of weeks per year you recycle = (___). The plastic bottles filling an average 20- to 25-gallon garbage can weigh about 6 pounds.

32. Calculate your lifestyle total:

32a. Add: 28d (___) + 29a (___) + 30a (___) = (___)

32b. If you compost, give yourself 2,000 gallons; if you don't compost, give yourself 0 gallons.

32c. Add: 31a (___) + 32b (___) = (___)

32d. Lifestyle total: Subtract: 32a (___) − 32c (___) = (___)

YOUR VIRTUAL WATER FOOTPRINT*	
CATEGORY	WATER FOOTPRINT
Food and Drink	
Home and Garden	
Lifestyle	
YEARLY TOTAL	

*The average American consumes 1,797 gallons per day (656,000 gallons per year).

conclusion

By now you should be as surprised as I am by how much water we use and can save. Notice that I wrote *"can* save" and not *"could* save."

We can do this. We can save gobs of water and avoid the crisis that will result if we run short of our precious water supply.

To be sure, it's simplistic and perhaps naïve to believe that a cup here or a liter there matters much in the scope of the world's water supply. But no one can argue that raising the level of consciousness about our water intake, waste, and conservation is anything but astute.

The areas feeding the Colorado River, which serves much of the American West, are experiencing their worst drought in 500 years. The Great Plains aquifer, the Ogallala, is disappearing, and in some places the groundwater is already gone. This is where one-fifth of the country's crops are grown. Filling it back up isn't an option. It would take up to 6,000 years to refill the reservoir. And the Great Lakes, as I've mentioned throughout this book, are suffering from myriad problems that deplete their freshwater supply, which collectively, might I remind you, is the largest surface source of freshwater in the world.

Now contemplate the pollution problems we are faced with. In recent years, violations of the 1972 Clean Water Act have risen steadily, according to the *New York Times*, which published an investigative series on the country's water problems in September 2009. Such violations result in water becoming either unusable or contaminated enough to pose health problems to those who drink it.

When we talk about health and wellness in this context, we are also talking about water conservation. It should be readily apparent that we can no longer afford to squander our water resources. Being "green" means caring about more than just global warming, it means caring about all of our natural resources.

In an ecosystem, things are connected. Air, water, and the food we eat are all connected to how we care for the planet. Climate change, pollution, and exploitation are all ways in which we exhibit our care—or lack of it. We can flip this around and promulgate healthier living and a healthier planet by taking positive actions: conserving, preserving, recycling, and cleaning up our waste.

As the *New York Times* wrote regarding the water supply in the United States: Today the nation's water does not meet the cleanliness standards set out by public health goals, and enforcement of water pollution laws is unacceptably low. Ditto in the rest of the world.

There are more people on the planet every day, using more natural resources, goods, and services. All of these things take water. Look back at the pages you've just read: The total water content of all the things listed in this book is about one-68 billionth of what we use. In other words, we use trillions of gallons of water a day. There are 10,460 cubic miles of freshwater available on the planet each year. To get your brain around that, it's enough to cover the entire surface of the Earth in 3 feet of water, and we are like little kids jumping up and down in that puddle, especially we the people of the United States.

We, along with Canada, Australia, Argentina, and Thailand, export—yes, export—a lot of the water that falls on our lands, collects underground, or melts into a liquid form we can use. We're among the top "virtual water" exporters of the world because of all the water we use to produce the goods we ship around the planet. Meanwhile, it's getting pretty arid around here.

If we can begin to look at the source of what we buy, use, or grow in the context of the water needed, we can begin to think about growing or making things in smarter ways, ways that use less water.

This type of water trade plays into the concept of virtual water. If we can figure out on a local level how to make choices that save water, we can apply the same thinking on a national level, as well as on an international level. Making the water footprints of products and entire nations transparent gives us the opportunity to create a more equitable world of water by balancing the trade between water-scarce and water-rich countries. Yet, this balanced approach hasn't been given the attention it needs in world trade agreements.

Actions speak a lot louder than words. When it comes to water, we have to begin to shout. Some of us, of course, will have to shout louder (i.e., save more water) to make up for those quiet people who refuse to change until they are forced to. This book can act as a bullhorn. It can help people "get it"; many don't when it comes to water use.

On a recent afternoon, a camera crew from a local news organization followed around a "water cop"—an official from the City of Los Angeles. This cop was handing out tickets and fines to people. Their crime? Watering their lawns on a day when rationing was enforced. Behind one gate, a man threw up his hands in surrender; the hose he had been holding fell to the ground, spouting water onto his lawn. Others had no idea there was even a rationing law. And there were, of course, the obstinate, "I don't care if you fine me" people who refused to adhere to the ban, which is only lifted 2 days per week. The television news story was light-hearted and serious at the same time. I hope the same can be used to describe this book. I don't think draconian enforcement is the answer to our water problem. I believe coaxed enlighten-ment is the cure. (Still, water laws are a pretty good means of education.)

Jared Diamond asks in his great book *Collapse* what the peo-ple of Easter Island must have been thinking when they chopped down the last tree. We're sort of at that same place in history now when it comes to water. Let's hope that people in the future, who will be living on water-rationed diets, don't ask the same of us.

"They flooded crop fields? They wasted half their food sup-ply? They overwatered their lawns? Wait—they watered their lawns!? They let *how* much water go to waste? They didn't think about water?"

I, for one, don't want to go down in infamy. Don't get me wrong, I don't want to hydrate off a drip-tube either. But can I watch my use? Sure. Can I shop a little smarter? Absolutely. Can I figure out ways to conserve and recapture water? Yup. It's not too much trouble, according to what I've learned while writing this book. (Trimming the water I use for my plants by diverting it from my tub or gutter seems like a no-brainer, for example.) In fact, there's a lot in this book that I can do with the knowledge I've gained and a little effort.

Some of my favorites: The one about licorice: It's 50 times as sweet as sugar, so you need less of it. Also, licorice actually improves water tables and helps other crops grow because it's a soil enhancer. The one about onions: Slicing an onion releases a gas that attaches to sensory neurons in the eye to make it tear. Don't cut the root of the onion, because this releases the most gas. Also, the finer the knife, the less cell damage and the less gas produced. The one about pasta: It takes only 1½ quarts of water to cook a pound of pasta, whereas most instructions say it takes 4 to 6 quarts. And the one that actually sparked my deep dive into researching this book: Measure your coffee-making water more accurately and you won't leave behind an extra cup at the bottom of the coffeepot; this small effort could potentially save 3 billion gallons of water in the United States alone—enough to save all the people who die every year from lack of access to freshwater.

I always find myself at the conclusion of my books, wondering why we don't know better, why we aren't told better, and why we don't act better. And I find myself running out of excuses.

A rather well-respected study of people's habits found that we change most when we know what other people have done or are doing to change. According to this study, numbers and simple solutions didn't affect habits as much as trying not only to keep up with the Joneses but also to beat them!

Some cities and towns have begun to put this thesis to work by showing people on their utility bills how their water use compares to their neighbors'. In other words, these utilities are using community pressure to sway behavior. It works. That's why the calculator provided in Chapter 14 is so important. It allows you to tally your use and compare it to the national average. If all goes according to plan, we'll be able to share resources on *The Green Blue Book*'s Web site to show even more exact comparisons.

But those comparisons are there for us individuals to get competitive with. My hope is that countries will join in the competition, too. If we can get governments to include water use in trade agreements, and if we can get businesses to wise up, we can solve the water crisis.

A movement begins with one action. And so we end with that beginning in mind.

for more information and resources

General Information and News on Water Issues

US Geological Survey, Water Resources of the United States
http://water.usgs.gov

Food and Water Watch
www.foodandwaterwatch.org/water

US Environmental Protection Agency, WaterSense
www.epa.gov/watersense

National Academy of Sciences, Water Information Center
http://water.nationalacademies.org

World Health Organization
www.who.int/topics/water/en

World Water Council
www.worldwatercouncil.org

Food and Agriculture Organization, Water Development and Management Unit
www.fao.org/nr/water

Virtual Water

Water Footprint Network
www.waterfootprint.org

Stockholm International Water Institute
www.siwi.org

Water Conservation Solutions

Water—Use It Wisely
www.wateruseitwisely.com

H_2ouse
www.h2ouse.org

H_2O Conserve
www.h2oconserve.org

Greywater
www.greywater.com

Water Quality

American Water Works Association, DrinkTap.org
www.drinktap.org

Water Quality Association
www.wqa.org

US Environmental Protection Agency, Ground Water and Drinking Water
www.epa.gov/OGWDW

Water Policy and Management

U.S. Bureau of Land Management, BLM Water Rights Policy
www.blm.gov/nstc/WaterLaws/blmwaterpolicy.html

Freshwater Action Network
www.freshwateraction.net

Water Research, Science, and Technology

Water Research Foundation
www.waterresearchfoundation.org

WateReuse Foundation
www.watereuse.org/foundation

Pacific Institute, The World's Water
www.worldwater.org

To Take Action

Clean Water Fund
www.cleanwaterfund.org

Lifewater International
www.lifewater.org

The Bonneville Environmental Foundation
www.b-e-f.org

Water for People
www.waterforpeople.org

Water Missions International
www.watermissions.org

Water Education Foundation
www.watereducation.org

Waterkeeper Alliance
www.waterkeeper.org

references

All the items and actions in this book have been sourced. In this section there are both publication details and Web links for the research. The links are provided to accomplish two missions: (1) to direct readers to the source material and (2) to help them find additional information about a particular item. Please see pages 131-132 for additional resources and more general information on water-related issues.

Unless otherwise noted, the base data for virtual water content were derived with permission from the Water Footprint Network and Arjen Hoekstra of the UNESCO-IHE Institute for Water Education and the University of Twente, the Netherlands. Data have been converted into US figures and, where possible, US virtual water content values were used instead of global averages. The exception to this is the use of the industrial product conversion based on a global average of 80 liters per US$1. Global assumptions are specifically written out as such.

It should be further noted that the assumptions and references vary by process, geography, and other determinants.

As stated, all items have been sourced and vetted as stringently as possible, but during the process of editing and fact-checking, sources may have gone amiss, awry, or out-of-date; we're making room on *The Green Blue Book*'s Web site to post corrections and amplifications.

The sources cited in this book are solely responsible for the veracity of the information provided by and attributed to them.

I calculated certain virtual water content figures myself. These calculations are shown below, where possible and appropriate, to show the methodology and rationale behind each conclusion.

My mission was and is to provide transparency in the data, narrative, and information provided. I strove to make things clear—as clear as water.

SECTION ONE

Introduction

Two-thirds of the global population will face water shortages by 2025:
Barlow, Maude. 2008. *Blue Covenant*. New York: New Press, p. 3
Freshwater makes up < 1 percent of the total volume of water on Earth:
http://ga.water.usgs.gov/edu/earthwherewater.html
Beef, beer, wine, coffee, tea:
Chapagain, A.K., and Hoekstra, A.Y. 2004. *Water footprints of nations. Volume 2: Appendices.* Value of Water Research Report Series No. 16. Delft, The Netherlands: UNESCO-IHE Institute for Water Education. http://www.waterfootprint.org/Reports/Report16Vol2.pdf
Car: 39,000 gallons; bicycle: 480 gallons:
http://www.nypirg.org/ENVIRO/water/facts.html
Average person uses more than 656,000 gallons of water per year:
Hoekstra, A.Y., and Chapagain, A.K. 2007. *Water footprints of nations: Water use by people as a function of their consumption pattern.* Water Resources Management 21(1):35–48. http://www.waterfootprint.org/Reports/Hoekstra_and_Chapagain_2007.pdf
70 percent or more of residential use is for our lawns:
Snitow, A., and Kostigen, Thomas M. 2008. Thirst response. *Best Life* December 23. http://www.mhbestlife.com/cms/publish/health/Natural-Resource-Conservation.php
California is rationing its water:
http://www.ens-newswire.com/ens/feb2009/2009-02-27-093.asp
Arizona imports water:
Barlow, Maude. 2008. *Blue Covenant*. New York: New Press, p. 4.
Georgia almost ran out of water:
http://www.muninetguide.com/articles/atlanta-water-249.php.
Texas is the driest region in country:
http://www.dallasnews.com/sharedcontent/dws/dn/latestnews/stories/030109dntswdrought.15a536b.html

Great Lakes fend off water poachers:
McCarthy, John. Ohio agreement to join Great Lakes water plan stalls again. Associated Press, May 30, 2008. http://www.whas11.com/sharedcontent/APStories/stories/D90VNB9OR.html

Chapter 1. Home

Introduction

The average household in America uses about 400 gallons of water per day: http://www.epa.gov/WaterSense/pubs/outdoor.htm

Bill Gates's mansion uses 4.7 million gallons of water per year: http://community.seattletimes.nwsource.com/archive/?date=20010427&slug=wateruse27m1

The average size of new homes in the United States is more than 2,500 square feet—780 square feet larger than homes built 3 decades ago: http://www.census.gov/const/www/highanncharac2008.html

Nineteenth-century average cost of indoor plumbing: http://jcgresidential.com/Documents/History%20of%20Plumbing%20in%20America.pdf

A person needs about 50 liters (13 gallons of water) per day to stay clean and healthy: http://rehydrate.org/water

Global population tripled while water consumption increased sevenfold: Barlow, Maude. 2008. *Blue Covenant*. New York: New Press, p. 3.

Bathroom

Brushing

Gallons per minute for a faucet: http://www1.eere.energy.gov/femp/pdfs/29267-6.3.pdf

5 gallons × 305 million people = 1.525 billion gallons per day

Average population of states is 6 million: http://en.wikipedia.org/wiki/List_of_U.S._states_by_population

1.525 billion ÷ 6 million people = 250 gallons

If each of these people uses 100 gallons per day, this is 2.5 days' worth of water.

Flushing

Toilets account for 26.7 percent of indoor water use. For a household of four that uses 400 gallons per day, this is 107 gallons per day: http://www.epa.gov/WaterSense/pubs/indoor.htm

If the average person flushes 6 times per day and saves ½ gallon per flush, this is 3 gallons saved per person, or 12 gallons saved per household; 12 gallons × 365 days = 4,380 gallons per year.

Showering

Average American showers for 8 minutes per day:

http://www.ncbuy.com/news/2001-07-19/1001877.html

Low-flow showerheads:

http://www.eartheasy.com/live_lowflow_aerators.htm

Baths

The average bathtub holds about 50 gallons of water, or 30 gallons more than taking an average shower:

$30 \times 365 = 11,000$ gallons

http://www.co.mchenry.il.us/departments/waterresources/pdfDocs/
waterusedbyhomes.pdf

The average above-ground pool holds between 10,000 and 12,000 gallons:

http://www.poolinfo.com/Pool-Volume.htm

Shaving

Assume shaving uses 8 gallons of water during a 5-minute shave. If filling the sink could cut this in half, then that is 90 gallons saved per month (with no shaving on weekends). Assume 107 million adult males in the United States. The savings would be 90 gallons $\times$ 12 months $\times$ 107 million males = 115.56 billion gallons per year.

http://www.census.gov/prod/2005pubs/06statab/pop.pdf

The Rose Bowl could hold 84,375,000 gallons:

115.56 billion $\div$ 84.375 million = 1,370 times

http://www.rosebowlstadium.com/RoseBowl_general-info.htm

Kitchen

Coffee Making

http://www.national-coffee-guide.com/Coffee%20Guide/Coffee%20Article.htm

220 million adults in the United States:

http://www.census.gov/prod/2005pubs/06statab/pop.pdf

48 percent drink coffee daily:

http://www.reuters.com/article/topNews/idUSN0326384920070303

220 million $\times$ 0.48 = 106 million

106 million $\times$ 1 cup $\times$ 365 days = 38.544 billion cups = 2.4 billion gallons

http://discovermagazine.com/2008/jun/28-everything-you-know-about-water-
conservation-is-wrong

Cooking

It's possible to cook a pound of pasta in just 1.5 quarts of water instead of 6, saving more than 1 gallon per pound. At 1 billion pounds consumed in the United States per year, the savings could be 1 billion gallons of water.

http://www.nytimes.com/2009/02/25/dining/25curi.html?pagewanted=1&_r=1

Drinking

http://www.snopes.com/medical/myths/8glasses.asp

Ice

105 million households × 1.5 cups = 9.84 million gallons

http://ga.water.usgs.gov/edu/watercyclesublimation.html

Waste

Garbage disposal:

5 gallons per day × 365 days = 1,825 gallons per year

Dishwashing

Average dishwasher uses 11 gallons per load:

http://www.energyguide.com/library/EnergyLibraryTopic.asp?bid=txu&prd=10&TID
=12817&SubjectID=7658

Allow four loads per week on average:

11 × 4 × 52 = 2,288 gallons

Laundry Room

http://factfinder.census.gov/servlet/SAFFFacts

Install a high-efficiency washing machine = 20 gallons saved:

http://www.save20gallons.org/tips.html

Appliances

Front-loaders use 27 gallons per load, top-loaders use 40 gallons per load. The
difference (13 gallons) could total 5,000 gallons for an average household. For
80 million washers, the savings could equal 400 billion gallons.

http://www.energyguide.com/library/EnergyLibraryTopic.asp?bid=txu&prd=10&TID
=12818&SubjectID=7658

Frequency

Wash full loads:

3.785 liters per gallon × 7,500 gallons = 28,390 liters

http://www.energyguide.com/library/EnergyLibraryTopic.asp?bid=txu&prd=10&TID
=12818&SubjectID=7658

http://www.fi.edu/guide/schutte/howmuch.html

Living Areas

Plants

http://www.gardenersgardening.com/indoorplants.html

http://www.voiceofsandiego.org/articles/2008/03/14/news/xeriscape031408.txt

Faucet Leaks

http://www.h2ouse.org/tour/details/element_action_contents.
cfm?elementID=1D4BABB7-8E4C-4524-98836EECCC5AEE08&actionID=F56F
50F2-34E3-4095-9A919C304D945B5F&roomID=8183044A-3219-48E2-
A965ACB77A568AC4

Toilet Leak Detection

http://www.h2ouse.org/tour/details/element_action_contents.
    cfm?elementID=5812B5A5-E0BE-4D14-A202C8DAE8CE491F&actionID=F56F5
    0F2-34E3-4095-9A919C304D945B5F&roomID=8183044A-3219-48E2-
    A965ACB77A568AC4

Aerators

http://www.lowesforpros.com/sites/default/files/Water%20Efficient%20Bath%20
    Solutions.pdf

Air Conditioners

Evaporative cooler:

http://ag.arizona.edu/pubs/consumer/az9145.pdf

http://www.dalby.qld.gov.au/Council/Docs/EvaporativeAircon.pdf

Tankless Heaters

http://www.energysavers.gov/your_home/water_heating/index.cfm/
    mytopic=12820

Meters

http://www.h2ouse.org/news/index.cfm

Pipes

Insulate water pipes:

http://www.recycleworks.org/greenbuilding/gbg_plumbing.html#sustainable_
    materials

http://articles.webraydian.com/article18223-Green_Plumbing___Save_Water_
    and_Energy_with_an_Efficient_Residential_Plumbing_Layout.html

Bottom Line

http://environment.about.com/od/greenlivingdesign/a/car_wash.htm

Chapter 2. Outdoors

Introduction

In some parts of the country—mostly the arid West—70 percent or more of
    residential water use occurs outdoors:

Mayer, Peter W., and Deoreo, William B., eds. 1999. *Residential end uses of water*.
    Denver: American Water Works Association.

120 gallons of water per day, on average, for things outside:

http://www.epa.gov/WaterSense/pubs/indoor.htm

Water use rights by states:

http://www.nytimes.com/2009/06/29/us/29rain.html

Outdoor water use accounts for 7 billion gallons per day, and half is wasted:

http://www.epa.gov/WaterSense/pubs/whatsnext.htm

Exteriors

Gutters

Gutter diverters and rain barrels:

http://www.wateruseitwisely.com/100-ways-to-conserve/outdoor-tips/how-to/
water-saving-toolbox/and-some-not-so-common-out.php

2,000 square feet of roof space that receives 1 inch of rainfall = 166 cubic feet =
about 1,200 gallons

Assume an average temperate-zone home receives 24 inches of rain per year, which
equals about 30,000 gallons of water—or a little less than 100 gallons per day:

http://www.wrcc.dri.edu/pcpn/us_precip.gif

Roofing

Water storage tanks in Texas:

http://bigcountryhomepage.com/content/fulltext/?cid=120921

http://wiki.answers.com/Q/How_many_gallons_of_rain_water_are_collected_at_
a_rate_of_one_inch_per_hour_on_a_ten_foot_by_twenty_foot_roof_at_
forty-five_degree_angle

Recapturing Systems

Gray water systems:

http://oasisdesign.net/greywater

http://www.bestsyndication.com/?q=081607_recylced-water-shortages-in-texas.htm

http://pa.photoshelter.com/c/green-stock-media/gallery/Wastewater-Recapture-
System-Residential-Shower/G0000GGoPwA1HShU

Lawn and Garden Areas

Maintenance

Cut grass no lower than 2 inches:

http://www.metrocouncil.org/environment/watersupply/conservationtoolbox_
residential.htm

Rogers, Elizabeth, and Kostigen, Thomas M. 2007. *The Green Book*. New York:
Three Rivers Press.

Snitow, A., and Kostigen, Thomas M. 2008. Thirst response. *Best Life* December 23.
http://www.mhbestlife.com/cms/publish/health/Natural-Resource-
Conservation.php.

Timers

Water before dawn to prevent losses due to evaporation:

http://www.epa.gov/WaterSense/pubs/water_efficient.htm

http://www.mass.gov/dep/recycle/reduce/dtg.htm

http://www.saskh2o.ca/PDF/EPB54CWaterSavingOutsidetheHome.pdf

Sprinklers

The average sprinkler sprays 240 gallons per hour:

http://www.libertylake.org/water_conservation.htm

http://www.expertvillage.com/video/12170_sprinklers-water-conservation.htm

Hoses

Water flows from an unrestricted garden hose at 12 gallons per minute:
http://www.thewatergeeks.com/Water-Saving-Hose-Nozzle-p-28.html
http://www.nextag.com/12-gallon-aquarium/products-html

Fountains

Water fountains—best dimensions:
http://ag.arizona.edu/AZWATER/arroyo/073fount.html

Plants, Xeriscaping, and Maintenance

Replace the lawn with drought-resistant plants:
http://www.voiceofsandiego.org/articles/2008/03/14/news/
    xeriscape031408.txt

Soils

Soil types and drainage:
http://www.ehow.com/how_16887_determine-much-water.html
Compost improves soil water storage, which saves water:
http://www.dmww.com/UsingWaterWisely/ReduceYourWaterNeedsWithCompost.
    pdf
http://chetday.com/overwatering.htm
Fertilizers:
http://www.dmww.com/UsingWaterWisely/ReduceYourWaterNeedsWithCompost.
    pdf

Pool

The average aboveground pool holds between 10,000 and 12,000 gallons:
http://www.poolinfo.com/Pool-Volume.htm

Types

Water-neutral pool:
http://www.spasavic.com.au/water-neutral-pool

Maintenance

Water levels, backwash filters, and splashing:
http://www.azwater.gov/AzDWR/StatewidePlanning/Conservation/Technologies/
    documents/pool_spas.pdf
http://www.barwonwater.vic.gov.au/emplibrary/00124C95-EB79-592D-
    7CB18OCD6B42FBCD.pdf?CFID=907030&CFTOKEN=58352032

Covers

http://www.inventright.com/media/1st-water-conservation—half-pool-cover-
    handler-44.html
http://www.netsquared.org/blog/solarfactory/pool-water-conservation-using-12-
    pool-covers

Windbreaks

http://www.azwater.gov/AzDWR/StatewidePlanning/Conservation/Technologies/
    documents/pool_spas.pdf

Drainage

http://www.azwater.gov/AzDWR/StatewidePlanning/Conservation/Technologies/
    documents/pool_spas.pdf

http://www.enewsbuilder.net/watercon/e_article000258807.cfm

Bottom Line

http://www.epa.gov/watersense/fixaleak

Chapter 3. Work

Introduction

Five beverage companies use as much water as everyone on the planet:

As reported by Devinder Sharma (http://www.culturechange.org/cms/index.
    php?option=com_content&task=view&id=402&Itemid=64), *The Economist* on
    August 27, 2008, stated: "Five big food and beverage companies—Nestle,
    Unilever, Coca-Cola, Anheuser-Busch and Danone—consume almost 575 billion
    litres of water a year, enough to satisfy the daily water needs of every person
    on the planet."

20 percent of global water supplies support industry; 9 percent are for domestic
    use:

http://www.fao.org/nr/water/aquastat/regions/lac/index4.stm

Time spent at work:

http://www.bls.gov/tus/charts

2 billion gallons saved: Assume Department of Labor's June 2007 total US
    workforce count of 146.14 million × 15 gallons (approximately 3 flushes at 5
    gallons + cups)

There are 27.2 million businesses in the United States:

http://web.sba.gov/faqs/faqIndexAll.cfm?areaid=24

27.2 million businesses × 1 cup × 250 days per year = 6.8 billion cups of water

The United States annually uses 4 million tons of copy paper, and the average
    office worker uses 10,000 sheets (100 pounds) per year:

http://eetd.lbl.gov/paper/html/concept.htm

Paper

1 ton of virgin uncoated paper = 18.7535 gallons

http://www.techsoup.org/learningcenter/techplan/page5675.cfm

100 percent recycled paper uses 5.1625 gallons per pound
http://www.edf.org/papercalculator

If the average office worker uses 100 pounds of copy paper per year and 33 percent of this is from recycled materials, the water footprint is 100 pounds × 0.67 × 18.7535 + 100 × 0.33 × 5.1625 gallons per pound = 1,426.85 gallons per office worker

4 million tons of copy paper a year is:

4 million × 0.33 × 5.1625 + 4 million × 0.67 × 18.7535 = 57,024,300 gallons per year for paper in the United States:
http://www.epa.gov/waste/conserve/materials/paper/faqs.htm

Break Room

Coffee

It takes 37 gallons (590 cups) of water to grow the beans to make 1 cup of coffee
http://www.rawstory.com/news/mochila/Scientist_who_invented_virtual_wate_03192008.html

http://www.census.gov/epcd/www/smallbus.html

Snitow, A., and Kostigen, Thomas M. 2008. Thirst response. *Best Life* December 23. http://www.mhbestlife.com/cms/publish/health/Natural-Resource-Conservation.php.

Bubblers

Assume a 3-ounce paper cup weighs 0.003 pounds. Assume nonrecycled paper: 18.7535 × 0.003 = 6.1888 gallons

Rogers, Elizabeth, and Kostigen, Thomas M. 2007. *The Green Book*. New York: Three Rivers Press.

Restroom

Sensors

Custodians find water left on in the restroom up to 10 times per week:
http://www.publicpolicycenter.hawaii.edu/documents/brief005.pdf

Infrared sensors save up to 70 percent of the water used per hand washing:
http://www.overstock.com/Home-Garden/EZ-Faucet-Automatic-Sensor-Faucet-Adaptor/2447903/product.html

Sinks

WaterSense faucets save 200 gallons per person per year, or 30 percent:
http://www.epa.gov/watersense/calculator/index.htm

http://www.epa.gov/watersense/pp/bathroom_faucets.htm

Dual Flush Toilets

Dual flush toilets save 67 percent of the water used for flushing:
http://www.greenbuildingsupply.com/Public/Energy-WaterConservation/WatersavingToilets/CaromaDualFlushToilet/index.cfm

Waterless Urinals

http://www.waterless.com/index.php?option=com_content&task=view&id=17&Itemid=44#

Meetings

Cups

1 liter = 4.22675282 cups

Rogers, Elizabeth, and Kostigen, Thomas M. 2007. *The Green Book*. New York: Three Rivers Press.

Bottles

It takes 3 liters to make one 1-liter plastic bottle

http://www.pacinst.org/topics/water_and_sustainability/bottled_water/bottled_water_and_energy.html

Allow 146 million workers:

www.census.gov

1 liter = 4.22675282 cups

Waste

50 percent of food is wasted:

http://www.foodproductiondaily.com/Supply-Chain/Half-of-US-food-goes-to-waste

http://www.waterfootprint.org/?page=files/productgallery&product=chicken

http://www.epa.gov/osw/education/quest/pdfs/unit1/chap3/u1-3_solidwaste.pdf

Landscaping

Design

There are 705,000 office buildings in the United States:

http://www.eia.doe.gov/emeu/consumptionbriefs/cbecs/pbawebsite/office/office_howlarge.htm

Estimate is based on 7,000 square feet × 700,000 buildings:

An acre is a unit of area; 640 acres = 1 square mile. Hence, an acre is 1/640th of a square mile. Since a square mile contains 27,878,400 square feet, an acre is 27,878,400 ÷ 640, which is 43,560 square feet.

Maintenance

Lawns in the United States use about 652,000 gallons of water per year per acre of lawn:

http://www.news.colostate.edu/Release/508

Smart Water System

Irrigation weather sensor:

http://earth2tech.com/2009/01/13/malls-get-smart-with-water-management

Chapter 4. Sports

Introduction

Nearly 150 million sports spectators:
http://www.allcountries.org/uscensus/433_selected_spectator_sports.html
http://www.bls.gov/spotlight/2008/sports
http://www.bls.gov/spotlight/2008/sports/data.htm#chart09
Green stadium:
http://www.environmentalleader.com/2009/06/01/new-meadowlands-stadium-ups-ante-for-green-sports-venues
16 percent of Americans age 15 years or older (about 245 million) participate in sports or exercise on a given day, which is about 40 million Americans:
http://www.bls.gov/spotlight/2008/sports
http://factfinder.census.gov/servlet/DTTable?_bm=y&-state=dt&-ds_name=
   PEP_2008_EST&-mt_name=PEP_2008_EST_G2008_T006_2008&-
   redoLog=true&-_caller=geoselect&-geo_id=01000US&-geo_id=NBSP&-
   format=&-_lang=en

Summer

Swimming
9 million pools each saving ½ cup = 4.5 million cups = 281,250 gallons
The average in-ground pool holds approximately 20,000 gallons:
http://www.poolinfo.com/Pool-Volume.htm
Golf
There are 23,000 golf courses in the United States
http://www.worldgolf.com/courses/unitedstates/usa.html
50 million gallons per year for the average golf course × 23,000 golf courses in the United States ÷ 365 days per year = 3.15 billion gallons of water per day:
http://www.nytimes.com/2009/08/06/science/earth/06golf.html
http://www.worldgolf.com/courses/unitedstates/usa.html
Golf courses get water conservation of 50 percent by changing grass type:
http://www.usga.org/course_care/articles/environment/water/Water-Conservation-on-Golf-Courses
Baseball
73 million fans × 5 gallons per flush; 365 million gallons saved:
http://www.econlib.org/library/Enc/Sports.html
12.8 million people play softball recreationally:
http://www.snewsnet.com/cgi-bin/snews/14805.html
Running
38 million people run for exercise:
http://fitnessbusinesspro.com/news/rise_runner_numbers

Tennis

Water tennis courts at night:
http://www.addvantageuspta.com/(S(aftxkyu0rrubxwu32qwrwp45))/default.aspx/
    act/newsletter.aspx/category/ADDvantage/menuitemid/344/MenuGroup/
    ADD-depts/NewsLetterID/1000.htm
It takes 400,000 liters of water to maintain a single clay court:
http://www.smartwater.com.au/projects/round2/heatherdale/Documents/
    HeatherdaleTC_ProjectReport.pdf

Hydration

http://www.npr.org/templates/story/story.php?storyId=5630821
Institute of Medicine: "Drink 20-24 fl oz water for every 1 lb lost." http://
    sportsmedicine.about.com/od/hydrationandfluid/a/ProperHydration.htm
http://ga.water.usgs.gov/edu/propertyyou.html

Time of Day

Average American showers for 8 minutes:
http://www.ncbuy.com/news/2001-07-19/1001877.html
Research suggests that individuals who exercise in the morning tend to exercise
    the most consistently:
http://www.webmd.com/fitness-exercise/features/whats-the-best-time-to-exercise
10 percent of Americans shower twice a day:
http://findarticles.com/p/articles/mi_m1571/is_6_16/ai_59585379/pg_2
http://www.census.gov/Press-Release/www/releases/archives/population/001703.
    html
About two-thirds of people exercise in the afternoon:
http://www.bls.gov/spotlight/2008/sports/data.htm#chart09
306 million × 10 percent = 30.6 million × 20 gallons = 612 million gallons

Surface

Water-resistant court tiles:
http://www.basketball-goals.com/search-basketball-court-resources-H.htm

Winter

Skating

Hockey rink statistics:
http://www.falmouthicearena.com/funfacts.htm

Skiing

800,000 liters = 10 percent of a ski resort's snow-making needs of 8 million
    liters:
http://www.betterworldclub.com/travel/ski.htm

Gymnasiums

Why choose a drinking fountain?
http://www.articlesbase.com/advice-articles/why-choose-a-drinking-water-
    fountain-238094.html

41.5 million health club members in the United States:
http://cms.ihrsa.org/index.cfm?fuseaction=page.viewPage.cfm&pageId=19547

Chapter 5. Travel

Introduction

30 million Americans travel abroad each year:
http://tinet.ita.doc.gov/cat/f-2006-101-002.html
Luxury hotels use 475 gallons per day per room:
http://www.zerowaste.org/publications/GREEN_HO.PDF
The average US household uses 400 gallons per day:
http://www.epa.gov/WaterSense/pubs/outdoor.htm
Leaks account for 11,000 gallons of water wasted in the home per year—enough to
    fill a backyard swimming pool:
http://www.epa.gov/watersense/pubs/fixleak.htm
Tourists use three times more water:
Batta, Ravinder N. 2000. *Tourism and the environment: A quest for sustainability.*
    New Delhi: Indus Publishing. p. 69.

Leaving Home

Shutoffs
Pipes bursting around the world:
http://www.istt.com/doks/pdf/a40_per_00.pdf
Timers
Rain sensors and efficient landscape irrigation savings:
http://sustainablechoices.stanford.edu/actions/in_the_home/efficientlandscaping.
    html
Water Heater
Water evaporation:
http://books.google.com/books?id=VWnxpAxp6TMC&pg=PA501&lpg=PA501&dq=w
    ater+evaporation+per+degree+of+temperature+rise&source=bl&ots=9CtPGU8
    Yfk&sig=DF3gXqk4yiMTS-QTXLxjckpP5lc&hl=en&ei=z9UaStn3A5KctgOL3azhC
    A&sa=X&oi=book_result&ct=result&resnum=10
http://www.fastwaterheater.com/traditional-water-heaters.asp
Plants
Watering indoor plants before leaving for vacation:
http://www.associatedcontent.com/article/37472/methods_for_watering_indoor_
    plants.html?cat=7

Airport

Bottles
More than 600 million airline passengers:
http://www.transtats.bts.gov

Airline passengers buying bottled water before going through security:
http://www.hotelfandb.com/reader-picks/07january-nestle-bottled-water-sales.asp
http://www.polarisinstitute.org/bottled_water_backlash_has_many_drinkers_
tapped_out
http://www.creditunionsonline.com/news/2009/Number-of-Airline-Passengers-
Declines-11.8-Percent.html

Security

Transportation Security Administration confiscates more than 13 million items per
year:
http://www.associatedcontent.com/article/1112486/confiscated_items_at_
airports_where.html?cat=47

Bathrooms

Automatic toilets in airports:
http://msande277.wordpress.com/2009/05/15/bathroom-musings-courtesy-of-
portland-airport
http://www.creditunionsonline.com/news/2009/Number-of-Airline-Passengers-
Declines-11.8-Percent.html

Hotel

Usage

Standard hotel rooms use up to 200 gallons of water per day:
http://www.4hoteliers.com/4hots_fshw.php?mwi=1889
http://books.google.com/books?id=6k_PFbfGAUIC&pg=PA237&lpg=PA237&dq=ho
tels,+water+conservation&source=bl&ots=MlvvmInFuE&sig=67kf8TsFriyXT7JN
qZ0IzfQnfQ8&hl=en&ei=ogwcStb1EpXstAPHvJiqDA&sa=X&oi=book_result&ct=
result&resnum=10#PPA241,M1
56 percent of hotel room water use is for showering:
Seneviratne, Mohan. 2007. *A practical approach to water conservation for commercial
and industrial facilities.* Amsterdam: Elsevier/Butterworth-Heinemann.
4.4 million guest rooms in the United States:
http://www.ahla.com/content.aspx?id=3448

Mini Bar

25 percent of 4.476 million guest rooms × 63.1 percent occupancy × 3 liters per
bottle = 533,000 gallons
http://www.ahla.com/content.aspx?id=3448

Room Service

48,000 hotels / 16 cups × 365 days = 1.095 million gallons
http://www.ahla.com/content.aspx?id=3448
http://www.allbusiness.com/services/business-services/4334901-1.html

Laundry

325,851 gallons in an acre-foot; 3,800 acre-feet used = 1,238,233,800 gallons:
http://www.pacinst.org/reports/las_vegas/LasVegas_Appendix%20E.pdf
http://www.sizes.com/units/acrefoot.htm

Spa

1,200 gallons every 16 days = 27,375 gallons:
http://www.sunbreeze.greatesc.com/water_talk.htm
http://www.servicemagic.com/article.show.Hot-Tub-Basics.8990.html

Destination

Touring

30 million Americans travel abroad each year:
http://tinet.ita.doc.gov/cat/f-2006-101-002.html

Traipsing

Dust speeds up snowmelt in the Rockies:
http://www.latimes.com/news/nationworld/nation/la-na-pink-snow24-
   2009may24,0,1077488.story
http://www.americanhiking.org/pressrelease.aspx?id=687&terms=75+million

Geography

Altitude:
http://www.worldtravelers.org/altitudeillness.asp

Time of Year

http://www.nbep.org/admin/Summertime%20Condition.pdf
http://cssrc.us/web/14/publications.aspx?id=
   4248&AspxAutoDetectCookieSupport=1

California's Drought:

http://cssrc.us/(X(1)A(_xHxv_FWygEkAAAAYmJhOGU4MGQtODUONyO0ZG
   VILTk2M2UtNDZIYzVkOGMzZDkxyZLOhovtt71FiejJwRst1WCfZ7s1))/web/14/
   publications.aspx?id=4248&AspxAutoDetectCookieSupport=1

SECTION TWO

Chapter 6. Foods and Beverages

Introduction

Agriculture uses 70 percent of freshwater:
http://agron.scijournals.org/cgi/content/abstract/101/3/477
Third World countries use almost entire freshwater supply to grow food:
http://earthtrends.wri.org/updates/node/264
Virtual water management has saved 5 percent of water in agricultural production
   already:
http://www.worldwatercouncil.org/fileadmin/wwc/Library/Publications_and_
   reports/virtual_water_final_synthesis.pdf
United States imports more fresh produce than it grows:
http://www.foodandwaterwatch.org/food/pubs/reports/the-poisoned-fruit-of-
   american-trade-policy

Food labeling of country of origin:
http://www.ams.usda.gov/AMSv1.0/cool
Half of food wasted:
http://www.foodnavigator-usa.com/Financial-Industry/US-wastes-half-its-food

Fruits

Chapagain, A.K., and Hoekstra, A.Y. 2004. *Water footprints of nations. Volume 2: Appendices.* Value of Water Research Report Series No. 16. Delft, The Netherlands: UNESCO-IHE Institute for Water Education. http://www.waterfootprint.org/Reports/Report16Vol2.pdf.

Apricots

Fresh

½ cup dried = 1 cup fresh = 1 serving:
http://www.dietbites.com/Easy-Diet/fruit-group-serving-sizes.html
For cut-up fruits, 7 cotton-ball-size chunks = 1 serving
http://www.goaskalice.columbia.edu/0715.html
10 medium-size apricots = 1 pound:
http://homecooking.about.com/od/foodequivalents/a/apricotequiv.htm
3 per cup:
http://seattletimes.nwsource.com/html/foodwine/2003726592_platkin30.html?syndication=rss
66 gallons of water per pound (http://www.waterfootprint.org/?page=files/NationalStatistics, Appendix IV) ÷ 10 apricots per pound = 6.6 gallons of water per apricot
6.6 gallons of water per apricot × 3 apricots per serving = 19.8 gallons for 1 cup = 1 serving.
Growing season for apricots: http://www.foodreference.com/html/artapricots.html

Dried

6 pieces = 40 grams = 1.4 ounces or 0.088 pounds for a serving
228 gallons of water per pound × 0.088 pounds per serving = 20 gallons of water per serving = 6 dried pieces
http://seattletimes.nwsource.com/html/foodwine/2003726592_platkin30.html?syndication=rss

Avocados

Average of 60 pounds from 150 fruits = 0.4 pound per fruit
0.4 × 106.52 gallons per pound = 42.6 gallons:
http://www.avocado.org/about/varieties

Bananas

437 cubic meters per ton of bananas for the United States = 52.37 gallons per pound
http://www.waterfootprint.org/?page=files/NationalStatistics, Appendix XVI
http://www.onlineconversion.com/volume.htm
Average weight of a banana: 100 grams:
http://www.botanical-online.com/platanos1angles.htm

Nearly all are imported and 1 pound = 3 bananas

http://www.fruitandveggieguru.com/Bananas.html?pccid=4&tabid=70&kw=
  Bananas

52.37 ÷ 3 = 17.5 gallons per banana

Grapefruit

http://www.oceanspray.com/news/events_2.aspx

http://www.innvista.com/health/foods/fruits/grapefr.htm

http://www.hort.purdue.edu/newcrop/morton/grapefruit.html

Kiwifruit

4.19 kiwis per pound, 1 kiwi per serving:

www.ers.usda.gov/data/fruitvegetablecosts/Kiwi.xls

0.25 pound per kiwi:

http://answers.yahoo.com/question/index?qid=20060926101545AAbmX57

512 cubic meters per ton for the United States:

http://www.waterfootprint.org/?page=files/NationalStatistics, Appendix XVI, p. 16.

http://en.wikipedia.org/wiki/Kiwifruit

Lemons

4 lemons per pound:

http://www.post-gazette.com/pg/09043/948477-34.stm

http://www.chow.com/ingredients/43

http://www.greenharvest.com.au/greennotes/Organic_Citrus_Care.html

http://meyerlemontree.com/watering.html

Limes

http://university.uog.edu/cals/people/PUBS/Lime/He8528.pdf

http://www.whfoods.com/genpage.php?tname=foodspice&dbid=27

Mangoes

Most mangoes are imported from Mexico, the Caribbean, Haiti, and South America.
  Some are commercially grown in Florida:

http://www.mamashealth.com/fruit/mango.asp

Mango facts about drought resistance and watering:

www.traditionaltree.org

Weight:

http://wiki.answers.com/Q/How_much_does_a_mango_seed_weigh

200 grams = 0.44 pound

186.1 gallons per pound × 0.44 pound per mango = 81.88 gallons

Melons

http://en.wikipedia.org/wiki/Melon

http://www.whfoods.com/genpage.php?tname=foodspice&dbid=17

Peaches and Nectarines

Canning peaches:

http://www.dinnerplanner.com/canning_peaches.htm

Peaches' and nectarines' average weight: 0.25 to 1 pound each:
http://www.peachesandpups.com/about.html
44.58 gallons per pound × 4 peaches per pound (at the smallest) = 11.145
    conservative estimate:
Funt, Richard C., and Schmittgen, Mark C. 2003. Evaluation of peach cultivars:
    1996-2003 yield, bloom, efficiency: Final summary, recommendations.
http://newfarm.osu.edu/crops/documents/EvaluationofPeachCultivars1996-2003.doc

Pears
http://www.sallybernstein.com/food/columns/ferray_fiszer/pears.htm
http://en.wikipedia.org/wiki/Pear

Pepper (Black)
http://en.wikipedia.org/wiki/Black_pepper

Pineapples
Range: 2 to 6 pounds, up to 20 pounds:
http://www.answers.com/topic/pineapple
Most common variety available in the United States is 5 to 6 pounds:
http://en.wikipedia.org/wiki/Pineapple
4 pounds per pineapple × 8.63 = 34.52
Bromelain:
    http://www.hungrymonster.com/FoodFacts/Food_Facts.cfm?Phrase_
    vch=Pineapples&fid=5924
Pineapple juice is mostly water, so it has a similar density; so use 8.35 pounds per
    gallon:
http://wiki.answers.com/Q/How_much_does_one_gallon_of_water_weigh

Plums
Average plum's weight is 109.9 grams = 0.24 pound:
http://www.freepatentsonline.com/PP13506.html
498 cubic meters per ton for the United States × 264.17 × 0.00045 = 59.67
59.67 gallons per pound × 0.24 pound per plum = 14.3 gallons per plum:
http://www.waterfootprint.org/?page=files/NationalStatistics, Appendix XVI, p. 10.
Growing season:
http://www.eatcaliforniafruit.com/csppn/stone-fruit-101/characteristics.asp
http://commhum.mccneb.edu/fstdatabase/HTM_files/Fruit/
    green%20gage%20plum.html

Nuts

Almonds
All almonds are grown in California; California grows 85 percent of the world's
    almonds:
http://www.ers.usda.gov/Briefing/FruitandTreeNuts/background.htm
1 cup of almonds (whole) = 8 ounces (0.5 pound):
http://www.almondboard.com/InTheNews/Documents/Tipsheet.pdf

518.4 gallons of water per pound × 0.5 pound per cup = 259.2 gallons:
http://www.veg-world.com/articles/cups.htm

Coconuts

1,189 cubic meters per ton in the United States for a coconut without a husk; that converts to 142.47 gallons per pound:

http://www.waterfootprint.org/?page=files/NationalStatistics, Appendix XVI, p. 12.

Average coconut's weight is 1.2 to 2 kilograms (use 1.6) with 36 percent of the husk. 1.6 — 0.576 (36 percent) = 1.024 kilograms = 2.25 pounds for the average coconut

2.25 × 142.47 = 320.6 gallons per coconut

Berries

Blueberries

Blueberries: 1 cup = 148 grams = 1 serving = 0.326 pound:
http://www.nutritiondata.com/facts/fruits-and-fruit-juices/1851/2

Blueberry virtual water content (42.4 gallons per pound):

Chapagain, A.K., and Hoekstra, A.Y. 2004. *Water footprints of nations. Volume 2: Appendices*. Value of Water Research Report Series No. 16. Delft, The Netherlands: UNESCO-IHE Institute for Water Education. http://www.waterfootprint.org/Reports/Report16Vol2.pdf, Appendix XIII, p. 150.

42.4 × 0.326 = 13.8 gallons

Blueberries peak in July. Fairly easy bush to grow at home, bear fruit in 3rd year:
http://www.gardenersnet.com/fruit/blueberry.htm

Grapes

Cluster weighs about 0.5 pound:
http://www.nysaes.cornell.edu/hort/faculty/reisch/bulletin/table/tabletext3.html

29.6 gallons per pound × 0.5 pound per cluster = 14.8 gallons per cluster

1/2 cup dried = 1 cup fresh = 1 serving:
http://www.dietbites.com/Easy-Diet/fruit-group-serving-sizes.html

For cut-up fruits, 7 cotton-ball-size chunks = 1 serving:
http://www.goaskalice.columbia.edu/0715.html

Raisins

6 pieces = 40 grams = 1.4 ounces or 0.088 pound for a serving

228 gallons per pound × 0.088 pound per serving = 20 gallons per serving = 6 dried pieces:

http://seattletimes.nwsource.com/html/foodwine/2003726592_platkin30.html?syndication=rss

pH Maturity:
http://www.extension.org/faq/1198

https://www.extension.iastate.edu/NR/rdonlyres/A647BBD4-08D5-494B-A55B-680667E6C342/79055/WhenaretheGrapesRipepdf.pdf

http://www.enowines.com/MT/archives/000166.html

Raspberries

1 cup = 123 grams = 0.271 pound:

http://www.nutritiondata.com/facts/fruits-and-fruit-juices/2053/2

67.82 gallons per pound × 0.271 = 18.379

Strawberries

1 cup (halves) = 152 grams = 0.335 pound:

http://www.nutritiondata.com/facts/fruits-and-fruit-juices/2064/2

10.78 gallons per pound × 0.335 pound per cup = 3.6113

http://missourifamilies.org/features/nutritionarticles/harvesttohealth/
strawberries.htm

Strawberry farm locations:

http://www.nal.usda.gov/pgdic/Strawberry/ers/ers.htm

Strawberries in pots:

http://www.doityourself.com/stry/growstrawberries

Vegetables

Alfalfa

Chapagain, A.K., and Hoekstra, A.Y. 2004. *Water footprints of nations. Volume 2: Appendices.* Value of Water Research Report Series No. 16. Delft, The Netherlands: UNESCO-IHE Institute for Water Education. http://www.waterfootprint.org/Reports/Report16Vol2.pdf, US value

Broccoli

http://en.wikipedia.org/wiki/Broccoli

http://www.cheftalk.com/cooking_articles/Cooking_Vegetables/107-Broccoli.
html

Cabbage

2 pounds = 1 head of cabbage:

http://king.wsu.edu/gardening/documents/41FallandWinterVegetableGardening_
000.pdf

http://www.bellybytes.com/recipes/cabbage.shtml

http://www.fao.org/nr/water/cropinfo_cabbage.html

Carrots

http://www.waterfootprint.org/?page=files/NationalStatistics, Appendix XVI,
p. 9.

Celery

http://www.healthnews.com/nutrition-diet/healthy-celery-more-than-just-a-
crudite-3123.html

http://www.whfoods.com/genpage.php?tname=foodspice&dbid=14

http://en.wikipedia.org/wiki/Celery

Corn

108 gallons per pound:

http://www.waterfootprint.org/?page=files/NationalStatistics

Cucumbers
237 cubic meters per ton for the United States = 28.4 gallons per pound:
http://www.waterfootprint.org/?page=files/NationalStatistics, Appendix XVI, p. 9.
Garlic
3 grams average per clove; has health benefits, including lowering cholesterol:
http://recipes.howstuffworks.com/natural-weight-loss-food-garlic-ga.htm
3 grams = 0.0066 pound
31.035 gallons per pound × 0.0066 = 0.205
Garlic is a powerful antioxidant:
http://www.sciencedaily.com/releases/2009/01/090130154901.htm
Lettuce
87 cubic meters per ton = 10.43 gallons per pound
http://www.waterfootprint.org/?page=files/NationalStatistics, Appendix XVI, p. 9.
Mushrooms
95 to 100 percent humidity:
http://en.wikipedia.org/wiki/Fungiculture#cite_note-r10-0
Water needed for misting logs:
http://attra.ncat.org/attra-pub/mushroom.html
http://en.wikipedia.org/wiki/Agaricus_bisporus
Onions
2 to 5 pounds for some:
http://www.onestoppoppyshoppe.com/servlet/the-Onion-Seeds/Categories
Range from less than half a pound to a pound:
https://secure.recipezaar.com/bb/viewtopic.zsp?t=308682&sid=
65394ff684cbf2f8b06505522f4ba720
http://en.wikipedia.org/wiki/Onion
Peas
1 cup of peas = 4 ounces (1/4 pound):
http://www.veg-world.com/articles/cups.htm
40.62 gallons per pound × 0.25 pound per cup = 10.155 gallons
Growing areas:
http://www.tonytantillo.com/vegetables/peas.html
Peppers (Bell)
More water = sweeter:
http://www.ehow.com/how_2043373_grow-bell-peppers.html
Change color—from green to red—as they ripen:
http://www.motherearthnews.com/Organic-Gardening/2006-04-01/Growing-
Colorful-Bell-Peppers.aspx
Potatoes
106 cubic meters per ton = 12.7 gallons per pound:
http://www.waterfootprint.org/?page=files/NationalStatistics, Appendix XVI, p. 9.

Drought-resistant varieties:

http://docs.google.com/gview?a=v&q=cache:C3h1D7moBdoJ:www.cals.uidaho.edu/
potato/Research%26Extension/Topic/Irrigation/
ManagingAPotatoCropWithReducedWaterSupplies-
06.pdf+ranger+russet+more+drought+resistant&hl=en&gl=us

Spinach

http://www.waterfootprint.org/?page=files/NationalStatistics, Appendix XVI, p. 9,
US value

Squash

http://en.wikipedia.org/wiki/Squash_(plant)

Sugar

Sugar from cane takes 1,500 liters per kilogram = 180 gallons per pound = 11.25
gallons per ounce:

http://www.waterfootprint.org/?page=files/productgallery&product=sugar

1 teaspoon sugar = 4.2 grams = 0.15 ounce

$0.15 \times 180 \div 16 = 1.6875$ gallons per teaspoon cane sugar = 4.2 grams:

http://nutrition.about.com/od/askyournutritionist/f/gramconversion.htm

Tomatoes

8.38 gallons per pound, 16 ounces per pound = 0.524 gallon per ounce × 2.5-ounce
tomato = 1.309

Irrigation and crop yield:

http://www.fao.org/nr/water/cropinfo_tomato.html

http://plantanswers.tamu.edu/vegetables/tomato.html

Meats

Beef

1,581 gallons per pound:

http://www.waterfootprint.org/?page=files/NationalStatistics

1,581 per 16 ounces × 6-ounce steak or burger = 592 gallons

Chicken

http://www.waterfootprint.org/?page=files/productgallery&product=chicken

http://www.ers.usda.gov/Briefing/Poultry/Background.htm

3,900 liters per 1 kilogram = 468.3 gallons per pound (global average):

http://www.waterfootprint.org/?page=files/productgallery&product=chicken

Lamb

http://en.wikipedia.org/wiki/Domestic_sheep#As_food

http://agecoext.tamu.edu/resources/library/newsletters/ag-eco-news-
series/2009/february-9-2009-the-us-sheep-inventory-down-for-3rd-year-
after-two-years-of-herd-rebuilding-herd-in-texas-down-94-percent.html

Pork

Hoekstra, Arjen Y., and Chapagain, Ashok K. (2008) *Globalization of water: Sharing
the planet's freshwater resources.* Malden, MA: Blackwell Publishing. http://
www.waterfootprint.org/?page=files/NationalStatistics

Turkey
http://www.ers.usda.gov/Briefing/Poultry/Background.htm

Dairy

Butter
30,000 liters per kilogram = 3,602.34 gallons per pound:
http://www.artmeetsscience.de/downloads/Art_meets_Science_Flemming.pdf
http://www.treehugger.com/files/2009/06/from-lettuce-to-beef-whats-water-footprint-of-your-food.php
http://books.google.com/books?id=DIoZAAAAYAAJ&pg=PA539&lpg=PA539&dq=butter,+water+content&source=bl&ots=4zSL-kIxW3&sig=Lx_eChYTBWWgacLcd8inECfGam0&hl=en&ei=F7chSo37Ipb8tgPOr730Aw&sa=X&oi=book_result&ct=result&resnum=7#PPA539,M1

Cheese
Fresh cheese:
http://www.e-cookbooks.net/cheese.htm

Eggs
180.94 gallons per pound × 0.126 pound per egg = 22.8 gallons per egg
57 grams for a large egg = 0.126 pound:
http://en.wikipedia.org/wiki/Egg_(food)
Egg production around the country:
http://www.ers.usda.gov/Briefing/Poultry/Background.htm

Milk
8.5 pounds per gallon of milk:
http://www.midwestdairy.com/pages/faq.cfm?TREE_ID=511

Yogurt
96.82 gallons per pound, 16 ounces per pound = 6.05 × average-size yogurt (6 ounces) = approximately 37 gallons per yogurt
Health benefits:
http://www.cspinet.org/nah/yogurt.htm
http://www.webmd.com/diet/features/benefits-of-yogurt

Grains

Barley
Fourth largest grain crop in the world:
http://www.duluthport.com/spring98/barley.html
About 100 to 200 gallons per pound, depending on how it is processed:
http://en.wikipedia.org/wiki/Barley
http://www.barleyfoods.org/facts.html#q2
Barley production, dry lands, and irrigation:
http://www.grains.org/barley

Oats

½ cup is a serving:

http://www.answerfitness.com/172/oatmeal-oats-oat-bran-healthy-food-day

42 ounces per 15.5 cups = 2.71 ounces per cup or 1.35 ounces per serving

1.35 ounces per serving = 0.084 pound per serving

194 × 0.084 = 10.3 gallons per serving

Rice

Zygmunt, Joanne. 2007. *Hidden waters*. Waterwise.org, February. pp. 14–16. www.waterwise.org.uk/.../EmbeddedWater/hidden%20waters,%20waterwise,%20february%202007.pdf.

1 cup of rice weighs 0.44 lbs:

http://www.traditionaloven.com/conversions_of_measures/rice_amounts_converter.html

Long, short, jasmine facts:

http://www.ers.usda.gov/briefing/rice/background.htm

Rye

39.78 gallons per pound × 0.22 pound per cup = 8.8 gallons per cup for 1 loaf of bread = 26.25 gallons

23 grams = 1 ounce = 1 serving:

http://caloriecount.about.com/calories-bread-rye-i18060

Pumpernickel and black breads are pure rye:

http://en.wikipedia.org/wiki/Rye_bread

1 cup rye flour = 0.22 pound:

http://www.traditionaloven.com/conversions_of_measures/flour_volume_weight.html

8.8 gallons per cup

Wheat or White Flour

Slice of bread (30 grams) = 40 liters = 10.5 gallons:

Zygmunt, Joanne. 2007. *Hidden waters*. Waterwise.org, February. Table 3. www.waterwise.org.uk/.../EmbeddedWater/hidden%20waters,%20waterwise,%20february%202007.pdf

1 cup uncooked rice = about 2 cups cooked. Approximately 200 gallons per pound, 0.44 pound per cup, gives 100 gallons per cup of uncooked rice, or about 50 gallons for 1 cup of cooked rice

Wheat is lower water per calorie:

Zygmunt, Joanne. 2007. *Hidden waters*. Waterwise.org, February. p. 18, Figure 11. www.waterwise.org.uk/.../EmbeddedWater/hidden%20waters,%20waterwise,%20february%202007.pdf

12 percent of global water consumption:

Hoekstra, Arjen Y., and Chapagain, Ashok K. (2008) *Globalization of water: Sharing the planet's freshwater resources*. Malden, MA: Blackwell Publishing. http://www.waterfootprint.org/?page=files/NationalStatistics

1 pound flour = 3.36 cups:

http://www.traditionaloven.com/conversions_of_measures/flour_volume_weight.
html

Bread recipes take 2 to 4 cups per loaf.

Cereal

Cornflakes

http://www.ksgrains.com/corn/talkingpoints.html

18-ounce box of cornflakes contains 12.9 ounces of milled field corn, or 71.6
percent:

http://ncga.com/corn-and-food-prices-record-4-17-08

Corn is 58.6 gallons per pound = 3.7 gallons per ounce. An 18-ounce box of
cornflakes contains 65.925 gallons of water plus water for malt, sugar,
vitamins, processing, etc.:

http://www.peertrainer.com/DFcaloriecounterB.aspx?id=

28 grams is one serving = approximately 1 ounce. An 18-ounce box = 18 servings.

Flax

Better than fish and dosage:

http://www.revolutionhealth.com/healthy-living/food-nutrition/healing-foods/
foods-heart/flax-flaxseed-oil

Ground flax is $2 per pound:

http://www.flaxseedmd.com/flax-seed-oil-uses-ground.asp

General information on flax:

http://www.rootsweb.ancestry.com/~belghist/Flanders/Pages/flaxLinen.htm

http://www.jeffersoninstitute.org/pubs/flax.shtml

http://www.hort.purdue.edu/newcrop/afcm/flax.html

Granola

Healthy Homemade Granola:

http://healthyhabitscoach.wordpress.com/2008/05/23/healthy-granola-recipe

Dry ingredients:

5 cups rolled oats (or a combination of rolled grains) (32 gallons per cup = 160
gallons)

2 to 3 cups raw almonds or pecan halves (or other nuts) (258 gallons per cup =
516 gallons)

½ cup dried fruit (add after cooked)—date pieces, cranberries, cherries, apricots,
mango (Raisins are 12.5 gallons per cup; apricots are 20 gallons per ½ cup).

160 + 516 + 20 = 696 ÷ 10 = 69.6 gallons per cup

Makes 10 cups

Wheat Bran

http://en.wikipedia.org/wiki/Bran

Pizza and Pastas

Adalya M.M., and Hoekstra, A.Y. 2009. *The water needed to have Italians eat pasta and pizza.* Value of Water Research Report Series No. 36. May. http://www.waterfootprint.org/Reports/Report36-WaterFootprint-Pasta-Pizza.pdf.

Beans

Beans

http://growingtaste.com/vegetables/bean.shtml

Soybeans

Chapagain, A.K., and Hoekstra, A.Y. 2004. *Water footprints of nations. Volume 2: Appendices.* Value of Water Research Report Series No. 16. Delft, The Netherlands: UNESCO-IHE Institute for Water Education. http://www.waterfootprint.org/Reports/Report16Vol2.pdf, Appendix XVI, p. 12. US value

Beverages

Cans

210 million cans / 8 million gallons:

http://www.docstoc.com/docs/5798520/facts-about-water-pollution

http://www.ecy.wa.gov/programs/hwtr/govaward/winners96.html

Glass Bottles

Grams of water per grams of glass (51 percent less in glass: 4,532 grams versus 8,826 grams per bottle):

http://www.triplepundit.com/2007/03/askpablo-glass-vs-pet-bottles/

http://www.convertunits.com/from/grams/to/liter

http://www.container-recycling.org

Plastic Bottles

1 liter = 0.26 gallon:

http://www.pacinst.org/topics/water_and_sustainability/bottled_water/bottled_water_and_energy.html

30 grams of plastic for a 12-fluid-ounce bottle

http://www.triplepundit.com/2007/03/askpablo-glass-vs-pet-bottles

Wasted daily:

$60 \times 2.3 = 138$

30×294.2 grams of water per gram of PET = 8,826 grams of water = 2.3 gallons

http://www.container-recycling.org

Soda

2.3 gallons per 12 ounces = 0.192

0.192×67.6 ounces (2 liters = 67.6 ounces) = 12.96 for the bottle itself. Plus all the sugars in the soda:

http://online.wsj.com/article/SB123483638138996305.html

Coffee

140 liters of coffee:
http://www.waterfootprint.org/?page=files/productgallery&product=coffee
140 liters = 591.7 cups:
http://www.onlineconversion.com/volume.htm

Tea

2 to 3 grams per bag = 0.005 pound:
http://wiki.answers.com/Q/In_a_pound_of_loose_tea_how_many_tea_bags_does_
it_equal
Black tea: 1,103.0 gallons per pound × 0.005 pound = 5.5 gallons per cup:
Green tea: 284.2 gallons per pound × 0.005 pound = 1.4 gallons per cup:
http://www.waterfootprint.org/?page=files/NationalStatistics, Appendix XVI,
p. 10.

Juices

Apple Juice

http://www.waterfootprint.org/?page=files/NationalStatistics, Appendix, XVI, p. 13.
US value

Orange Juice

8.35 × virtual water count per pound to get gallon measure

Pineapple Juice

http://extension.osu.edu/~news/story.php?id=2081

Tomato Juice

8.35 × virtual water count per pound:
http://extension.osu.edu/~news/story.php?id=2081

Snacks

Candy

http://www.new-ag.info/08/02/focuson/focuson7.php
http://www.food-info.net/uk/products/sweets/liquorice.htm

Popcorn

108.1 gallons per pound for corn per 16 ounces per pound × 1.4 ounces per bag =
9.5 gallons

Potato Chips

185 liters = 50 gallons:
http://knol.google.com/k/andreas-kemper/virtual-water/8bgikaqot3ts/237#

Wines and Spirits

Beer

24 to 25 pounds of malt per barrel of beer
4.7 billion pounds malt requires 5.9 million pounds barley

Average American drinks 20 gallons per year
20 gallons of beer requires 21 pounds of barley:
http://www.ag.ndsu.nodak.edu/aginfo/barleypath/barley&beer.html
21 pounds barley × 100 gallons water per pound = 2,100 gallons of water per 20 gallons beer
105 gallons per 1 gallon beer
1 gallon per 3.785 liters × 105 = 397.425 gallons per liter × (1 liter / 1,000 ml) × (250 ml) = 99.3 gallons water for 1 glass of beer (250 ml)
79.7 pounds of malt per 1 pound barley
1 barrel of beer = 31 gallons:
http://en.wikipedia.org/wiki/Barrel
http://www.waterfootprint.org/?page=files/productgallery&product=beer

Gin

http://en.wikipedia.org/wiki/Gin
http://www.tastings.com/spirits/gin.html
http://www.ehow.com/how_5034110_identify-harvest-use-juniper-berries.html
301.5 liters for bottle and alcohol = 79.7 gallons
42.4 gallons of water per pound of blueberry × 8.35 gives 354.2 gallons water per gallon blueberry juice
Assuming 1 gallon of blueberry juice per gallon of gin:
354.2 gallons of water per gallon + 79.7 gallons of water per gallon = 433.9 gallons of water per gallon
433.9 ÷ 3.79 = 114.48 gallons of water per liter of gin

Red Wine

http://www.waterfootprint.org/?page=files/productgallery&product=wine
http://www.food-info.net/uk/products/wine/intro.htm
http://www.njskylands.com/fmwineries2.htm
http://ezinearticles.com/?Understanding-Basics-to-the-Best-Red-Wine&id=1129098
http://www.sfgate.com/cgi-bin/article.cgi?f=/c/a/2007/06/01/WIG8OQ1CII1.DTL
http://goosecross.com/2009/06/wine-definition-glossary
For Pinot Noir:
http://www.eveshamwood.com/Site/Our_Wines.html

Rum

http://en.wikipedia.org/wiki/Rum
http://en.wikipedia.org/wiki/Alcohol_by_volume
50 percent alcohol by volume (typically 40 to 50 percent)
http://www.ministryofrum.com/article_how_rum_is_made.php
1 liter bottle will have 0.5 liter water = 0.13 gallon
1.5 liters for the bottle = 0.4 gallon
Rum comes from sugar cane mostly:
http://www.ministryofrum.com/article_how_rum_is_made.php

10 pounds of sugar or molasses gives 1 gallon rum

http://answers.yahoo.com/question/index?qid=20071017195131AA9prW3

10 pounds of sugar cane per 1 gallon of rum × 12.3 gallons of water per pound = 123 gallons of water per 1 gallon of rum

123 gallons of water ÷ 3.785 liters per gallon = 32.5 gallons of water per liter of rum.

32.5 gallons + 0.13 gallon for dilution + 0.4 gallon for bottle = 33 gallons per liter of rum

Tequila

50 to 60 kilograms of agave fruit yields 7.1 liters of tequila

894.9 gallons per pound for agave fibers

Hoekstra, Arjen Y., and Chapagain, Ashok K. (2008) *Globalization of water: Sharing the planet's freshwater resources.* Malden, MA: Blackwell Publishing. http://www.waterfootprint.org/?page=files/NationalStatistics

http://www.ianchadwick.com/tequila/tasting.htm

http://en.wikipedia.org/wiki/Tequila

Vodka

300 liters for 1 liter + 1.5 liters for the bottle + 0.5 liter of water = 302, which is 79.8 gallons:

http://www.tasteoftx.com/spirits/vodka.html

Average 50 percent water by volume:

http://en.wikipedia.org/wiki/Vodka

Barley per 250 milliliters = 300 liters per liter:

http://www.waterfootprint.org/?page=files/productgallery&product=beer

1,300 liters per kilogram:

http://www.waterfootprint.org/?page=files/productgallery&product=barley

Whiskey

http://en.wikipedia.org/wiki/Single_malt_Scotch

222 pounds of grain for 80 to 87 liters of whiskey (use 80):

http://www.probrewer.com/resources/distilling/whiskey.php

222 pounds of grain ÷ 80 = 2.775 pounds of grain for 1 liter of whiskey

Barley is 100 to 200 gallons per pound; use 150 gallons

2.775 × 150 gallons gives 416.25 gallons of water to grow grain for 1 liter of whiskey

1.5 liters for the bottle

50 liters added to make the mash

50 + 1.5 = 51.5 liters = 13.6 gallons

416.25 + 13.6 = 429.9 gallons for 1 liter of whiskey

White Wine

http://www.extension.org/faq/1198

http://www.waterfootprint.org/?page=files/productgallery&product=wine

http://en.wikipedia.org/wiki/Annual_growth_cycle_of_grapevines

http://en.wikipedia.org/wiki/Chardonnay

Chapter 7. Clothing

Introduction

8.32 million tons of clothing and footwear enter the waste stream annually:
http://www.epa.gov/epawaste/nonhaz/municipal/pubs/msw07-rpt.pdf

8.32 million tons = 16,640 million pounds. Divided by 306 million people = 54
pounds per person

California's urban water demand in 2000 was 2.9 trillion gallons:
http://www.ppic.org/content/pubs/cep/EP_706EHEP.pdf

10,850 liters (2,866.27 gallons) total virtual water content per kilogram of cotton =
4,900 liters of blue water + 4,450 liters of green water (total = 2,470 gallons) +
1,500 liters (396.26 gallons) of dilution water

Growing cotton lint: 4,506 liters (2,247 gallons)

622 liters (165 gallons) of dilution water for fertilizers and pesticides per
kilogram

368 liters (97 gallons) from cotton to gray fabric

360 liters (95 gallons) of water used for bleaching, dyeing, and printing 1 kilogram
of cotton

136 liters (36 gallons) of water for finishing process for 1 kilogram of cotton

880 liters (232 gallons) of dilution water for wet processing and finishing per
kilogram

Organic cotton makes up 0.55 percent of global cotton production:
http://www.alertnet.org/thenews/newsdesk/IRIN/
c862400d524a8e38e9330d0e628ec886.htm

Cotton makes up 40 percent of all textiles:
http://www.waterfootprint.org/Reports/Report18.pdf

Clothing

Jackets

Types of leathers:
http://ezinearticles.com/?Stylish-Leather-Jackets-For-Women&id=3019454

10.4 pounds for jacket (with liner):
http://www.amazon.com/Tour-Master-Coaster-Leather-Jacket/dp/B001A3GX7K

Gives 4 pounds and some others around this weight; use 4 pounds.
http://www.amazon.com/Classic-Length-Zipper-Leather-Jacket/dp/B00194CI1U/
ref=sr_1_1?ie=UTF8&s=apparel&qid=1245456213&sr=8-1

16,600 liters per kilogram then 4,385.26 gallons per kilogram then 1,989.12 gallons
per pound × 4 pounds per jacket = 7,956.47 gallons per jacket:
http://www.waterfootprint.org/?page=files/productgallery&product=leather
(global average)

Jeans and Pants

450 million pairs of jeans sold in the United States per year

Average woman has 8 pairs of jeans:

http://www.onearth.org/article/how-green-are-your-jeans

Cotton farmers use 25 percent of all insecticides globally.

Toxic dyes for coloring and chemical treatments to create "worn" look leak into waterways and pollute freshwater systems.

"Stressed" and "softer" often means multiple washings of fabric:

http://www.onearth.org/article/how-green-are-your-jeans

Assume 1 pair of jeans weighs 1 kilogram (2.2 pounds)

10,850 liters then 2,866 gallons:

http://www.waterfootprint.org/Reports/Report18.pdf

Running Sneakers

Weight of running shoes is 10 to 15 ounces. Assume an average of 12.5 ounces:

http://www.runnersworld.com/article/0,7120,s6-240-400—12428-0,00.html

Assume 60 percent of weight is rubber: $0.6 \times 12.5 = 7.5$ ounces of rubber

13,058 liters per kilogram of natural rubber, then 3,449.56 gallons per kilogram, then 1,564.69 gallons per pound:

http://www.waterwise.org.uk/images/site/EmbeddedWater/hidden%20waters,%20waterwise,%20february%202007.pdf

Using natural rubber = (1,564.69 gallons per pound or 16 ounces) = 97.793 gallons per ounce $\times$ 7.5 ounces = 733.448 gallons per shoe

Assume 20 percent canvas = 2.5 ounces and assume cotton for canvas: (1,300.12 gallons of water per pound cotton) / 16 ounces = 81.25 gallons per ounce $\times$ 2.5 = 203.144 gallons per shoe

Assume 20 percent leather (tanned) = (1,989.12 gallons per pound of leather) $\div$ 16 ounces = 124.3125 gallons per ounce $\times$ 2.5 = 310.8 gallons per shoe

Total shoe: $733.448 + 203.144 + 310.8 = 1,247.39$ gallons per running shoe

Trees in plantations:

http://en.wikipedia.org/wiki/Rubber#Current_sources for rubber production/collection

New Balance manufactured in the United States:

http://dietfitnesshealth.com/new-balance-footwear-quality-sneakers-manufactured-in-the-usa

Vans manufactured in China:

http://www.allbusiness.com/north-america/united-states-california-metro-areas/253226-1.html

Upper, midsole, and outsole:

http://www.sneakerhead.com/nike-brand-technology.html

Shirts

Long-sleeved men's shirt has about 12 ounces of cotton

1,300.12 gallons per pound of cotton $\times$ (1 pound / 16 ounces) $\times$ 12 ounces = 975 gallons per shirt

Avoid Egyptian cotton because water footprint there is even higher: 100 percent irrigated compared to only 50 percent in the United States:

Chapagain, A.K., Hoekstra, A.Y., H.H.G. Savenije, and Gautam, R. 2005. *The water footprint of cotton consumption.* Value of Water Research Report Series No. 18. http://www.waterfootprint.org/Reports/Report18.pdf.

Shoes

Most leather is cowhide:

http://ezinearticles.com/?Some-Of-The-Most-Popular-Leather-Shoes-On-The-Market-Today&id=338379

8,000 liters per pair of bovine leather shoes then 2,113.38 gallons per pair:

http://www.waterwise.org.uk/images/site/EmbeddedWater/hidden%20waters,%20waterwise,%20february%202007.pdf

Sheep or lamb skin uses 6,100 liters per kilogram, which is almost ⅓ of the water footprint of bovine leather (16,600 liters per kilogram).

Basic facts:

http://en.wikipedia.org/wiki/Shoe

Suede is primarily lambskin:

http://en.wikipedia.org/wiki/Suede

Leather:

http://en.wikipedia.org/wiki/Leather

Socks

Weigh 2 to 4 ounces; average is 3 ounces:

http://www.fiber2yarn.com/catalog.php?maincat=Fiber

10,850 liters per kilogram of cotton = 1,300 gallons per pound of cotton = 243.77 gallons per 3-ounce pair of socks

Suits

Typical men's suit fabrics:

http://www.theguidetomenssuits.com/suit-fabric.html

3 pounds average weight:

Cotton: 1,300.12 gallons per pound × 3 pounds = 3,900 gallons per suit

Linen (flax fiber): 3,134 liters per kilogram then 375.54 gallons per pound × 3 pounds = 1,126.62 gallons per linen suit:

www.waterfootprint.org/Reports/ResearchData/Appendix%20XV.xls

Wool: 6,100 liters per kilogram then 730.94 gallons per pound × 3 pounds = 2,192.82 gallons

Wool dyeing uses 250 liters per kilogram, or 30 gallons per pound, so 2,282.82 gallons per wool suit:

http://www.p2pays.org/ref/10/09237.htm

Cashmere (goat-based): 4,000 liters per kilogram then 479.31 gallons per pound × 3 pounds = 1,437.92 gallons per pound of cashmere

Assume dyeing uses same amount of water as for wool. Total is 1,527.92:

http://www.waterfootprint.org/?page=files/productgallery&product=goatmeat

Calculations assume suit weighs approximately the same for each material. Wool is likely heavier. Virtual water for wool and cashmere are based on water footprint for sheep and goat meat, respectively.

Sweaters

Use numbers for cotton for dyeing and bleaching:

Actual sweater weight: 352.71 grams = 12.44 ounces

730.94 gallons per pound + 30 gallons per pound for dyeing $\times \frac{1}{16} \times$ 12.5 ounces = 594.45 gallons per sweater

T-Shirts

Assume a T-shirt weighs 7 ounces

1,300.12 gallons per pound $\times$ ($\frac{1}{16}$ ounce) $\times$ 7 ounces = 568.80 gallons

Hemp: 2,507 liters per kilogram, then 300 gallons per pound $\times$ (1 pound or 16 ounces) $\times$ 7 ounces = 131 gallons per T-shirt

Underwear

Cotton ladies' underwear: 30 grams a pair = 1.058 ounces

1,300.12 gallons per pound of cotton $\times$ (1 pound or 16 ounces) $\times$ 1.058 = 85.99 gallons for one pair of cotton underpants

Boxers: cotton: 88 grams = 3.1041 ounces $\times$ 1,300.12 $\div$ 16 = 252.23 gallons for one pair of cotton boxer shorts

Silk not good: labor and water:

http://www.oldandsold.com/articles04/textiles16.shtml

Chapter 8. Furnishings

Introduction

Average American home has 10,000 items in it:

Rogers, Elizabeth, and Kostigen, Thomas M. 2007. *The Green Book*. New York: Three Rivers Press. p. 65.

Americans spend $78.5 billion a year on furniture, 38 percent of it made from wood:

http://www.prlog.org/10011975-household-furniture-consumption-in-the-united-states-of-america-with-forecast-to-2015.html

20 to 40 percent of the world's timber supply comes from illegal logging:

http://www.europarl.europa.eu/news/expert/infopress_page/064-54121-111-04-17-911-20090421IPR54120-21-04-2009-2009-false/default_en.htm

Environmental Protection Agency: 8.8 million-ton pile of furniture at our landfills:

http://earth911.com/household/home-and-office-furniture/facts-about-office-furniture

Products

Beds

150 pounds for a queen-size organic cotton/wool blend and same weight for hemp:

http://earthsake.com/shopsite_sc/store/html/HempMattress.html

Assume 70 percent is the metal springs, 30 percent is cotton/wool or hemp:

150 × 0.7 = 105 pounds of metal

Steel in a 30-pound bicycle takes 480 gallons of water:
http://www.nypirg.org/ENVIRO/water/facts.html

480 gallons ÷ 30 pounds × 105 pounds = 1,680 gallons for springs

150 × 0.3 = 45 pounds of fill

Cotton/wool blend: assume 50 percent each (45 × 0.5) = 22.5

Cotton: 1,300.12 gallons per pound × 22.5 = 29,252.7 gallons

Wool: 956 gallons per pound × 22.5 = 21,510

Cotton/wool fill = 21,510 + 29,252.7 = 50,762.7 gallons for fill

Total cotton/wool mattress: 1,680 + 50,762.7 = 52,442.7 gallons

Hemp: 45 pounds × 508.7 gallons per pound = 22,891.5 gallons + 1,680 for metal = 24,571.5 gallons

Man-made: 1 pillow = 21 × 15 × 5 inches = 1,575 inches:

http://www.discount-pet-mall.com/pet-products/barbados-comfort-dog-bed.html

And takes 40 bottles = 3 liters × 40 = 120 liters = 31.7 gallons

Queen mattress is 80 × 62 × 12 inches = 59,520 inches

59,520 ÷ 1,575 = 37.8 pillows' worth × 31.7 gallons per pillow = 1,198.26 gallons for fill

1,198.26 + 1,680 metal frame = 2,878.3 gallons per mattress:

http://www.nypirg.org/ENVIRO/water/facts.html

http://www.uship.com/shipment/Queen-Mattress-Boxed/739867889

http://en.wikipedia.org/wiki/Polyurethane

Blankets

956 gallons per pound of wool (processed)

Blanket weighs 4 pounds:

http://www.tradekey.com/product_view/id/181318.htm

956 × 4 = 3,824 gallons per blanket = 6,336 square inches

For acrylic/poly blanket, 3,500 square inches weighs 3.15 pounds. That's about half the size of wool blanket, so double: 6.3 pounds

Fill for pillow with polyethylene terephthalate bottles = 18.6 ounces = 1.2 pounds and was 31.7 gallons

31.7 gallons ÷ 1.2 pounds × 6.3 pounds = 166.4 gallons

Chairs

Leather, dimensions 38.5 × 40 × 31 inches, approximately 35.53 square feet of surface area of leather with weight of 95 pounds:

http://www.amazon.com/Black-Leather-Recliner-Chair-Hugger/dp/B000L2DJ8M

35.53 square feet leather × 2.5 ounces per square foot = 88.83 ounces of leather = 5.6 pounds of leather = 1,989.1 gallons per pound × 5.6 = 11,093.6 gallons for leather

Total chair weight: 95 pounds − 5.6 pounds of leather = 90 pounds of wood/steel

Assume 70 percent of weight is steel

480 gallons ÷ 30 pounds × 63 = 1,008 gallons

11,093.6 + 1,008 = 12,101

Chest

Surface area of 4,883.9 square inches × 0.5 (assume 1/2-inch thickness) = 2,441.95 cubic inches

2,441.9 cubic inches = 1.413 cubic foot = 16.9356 board-feet × 5.4 = 91.5 gallons: http://www.walmart.com/catalog/product.do?product_id=11018053&findingMethod=rr

Couches

Typical dimensions: 76.5 × 34 × 34.5 inches:

http://www.overstock.com/Home-Garden/Uptown-Collection-Peat-Sofa/3911915/product.html

Add up surface area = 16,453.9 square inches = 114.3 square feet

Leather: 1 ounce is 1/64th of an inch in thickness

Side leather: One-half of a full cowhide, cut right up the backbone. Makes up 18 to 22 square feet of total surface area:

http://www.brettunsvillage.com/leather/leatherterminology.htm

Hide is 50 to 55 square feet:

http://www.rodenleather.com/faqs.html

Couch is about 2 hides:

2.5 to 3 ounces per square feet:

http://www.interiormall.com/cat/ncollections.asp?c1=Fabric&c2=Leather

114.3 × 2.5 = 285.75 ounces for couch = 17.9 pounds

16,600 liters per kilogram = 1,989.1 gallons per pound

http://www.waterfootprint.org/?page=files/productgallery&product=leather

1,989.1 × 17.9 = 35,604.89 gallons for leather

Dinnerware

Dinnerware set of 4:

http://www.crateandbarrel.com/family.aspx?c=4310&f=32247

Each place setting (dinner plate, salad plate, bowl, and mug) is about 4 to 5 pounds

Total set is 20 pounds = 9,071.9 grams

5.3 grams of water per gram of ceramic:

http://www.triplepundit.com/pages/ask-pablo-the-c.php

5.3 grams of water per gram of ceramic × 9,071.9 grams of ceramic = 48,081.1 grams of water

48,081.1 grams × 1 kilogram / 1,000 grams × 1 liter / 1 kilogram × 1 gallon / 3.785 liters = 12.7 gallons of water.

Glassware

16-piece set of eight 17-ounce glasses and eight 12-ounce glasses weighs 15.2 pounds = 6,894.6 grams

6,894.6 grams × 17.1 grams of water per gram per glass:

http://www.walmart.com/catalog/product.do?product_id=10749453

117,897 grams water

http://www.triplepundit.com/2007/03/askpablo-glass-vs-pet-bottles

117,897.6 grams of water × (1 gallon/3785.4 grams) = 31.14 gallons

About 2 gallons per glass between 12 and 17 ounces in size

2 gallons ÷ 14.5 ounces = 2.2 cups per ounce × 8 ounces = 17 cups for an 8-ounce glass

Pillows

Weighs 1.6 pounds (fill and cover):

http://www.allergybuyersclubshopping.com/dreamsacks-silk-pillows.html

Assume cover is same weight as T-shirt and made of cotton; T-shirt weighs 7 ounces.

1.6 pounds = 25.6 ounces — 7 ounces = 18.6 ounces of fill

Cotton cover = 75 percent of cotton T (25 percent for the sleeves removed) = 569 × 0.75 = 426.8 gallons

Fill: 18.6 ounces = 527.3 grams

80 grams of feathers from one goose, and one goose weighs about 4 to 6 kilograms:

http://www.fao.org/DOCREP/005/Y4359E/y4359e0b.htm#TopOfPage

Goose meat is 50 percent of weight of bird:

Raw boneless meat = 3 servings per pound, serving is 3 to 4 ounces. So for a 12-pound goose (196 ounces), get 12 × 3 = 36 servings × 3 ounces = 108 ounces of meat: 108 ÷ 196 = 0.5:

http://www.fsis.usda.gov/FACT_Sheets/Duck_&_Goose_from_Farm_to_Table/index.asp

Requires same water as a chicken:

4 kilograms = 8.8 pounds × 0.5 × (468.3 gallons of water per pound of chicken) = 2,060.5 gallons per goose, that gives 80 grams of feathers

527.3 grams of feathers required = 527.3 ÷ 80 = 6.6 geese

2,060.5 × 6.6 = 13,599.4 gallons of water required for fill.

13,599.4 + 426.8 (cover) = 14,026.3 gallons for pillow of goose down

Man-made fill: 40 plastic bottles for one pillow fill

Assume 1 liter bottles = 3 liters of water per bottle × 40 bottles = 120 liters = 31.7 gallons

Cover of cotton = 426.8

426.8 + 31.7= 458.5 gallons

Pots and Pans

Stainless steel 1-quart saucepan is 2 pounds 5 ounces = 37 ounces = 1,048.9 grams:

http://www.chefscatalog.com/product/93520-all-clad-stainless-sauce-pan.aspx

205 grams of water per gram of stainless steel × 1,048.9 grams = 215,024.5 grams:

http://www.triplepundit.com/2006/09/ask-pablo-the-coffee-mug-debacle

1 kilogram is 1,000 grams × 1 liter per 1 kilogram × 1 gallon / 3.785 liters × 215,024.5 = 56.8 gallons

Rugs

5 feet × 8.9 feet cotton and wool rug (80 percent wool, 20 percent cotton) at 9.3 pounds:

http://www.novica.com/itemdetail/index.cfm?pid=135573&AID=10391713&SID=tfc_
-_8_9_090826_2d94c8b803084e9fc2d7fd4dd6a74885

7.44 pounds wool × 956 gallons per pound = 7,112.64

1.86 pounds cotton × 1,300.12 gallons per pound = 2,418.22

Total = 9,530.9 gallons

9,530.9 ÷ 44.5 square feet = 214.2 gallons per square feet of rug

All wool:

5.5 pounds for a 27.7-square-foot rug

5.5 × 956 = 5,258 gallons

5,258 ÷ 27.7 = 189.8 gallons per square foot

Sheets

400-thread-count cotton sheet: 5 pounds 2 ounces = 5.125 pounds

1,000-thread-count cotton sheet: 6 pounds 14 ounces = 6.875 pounds

Cotton: 1,300.12 gallons per pound

1,300.12 × 5.125 = 6,663.115 gallons

1,300.12 × 6.875 = 8,938.325 gallons

Silverware

18/0 stainless steel, 8 five-piece settings and 5 serving utensils:

http://www.walmart.com/catalog/product.do?product_id=4634053#ProductDetail

7 pounds total weight = 3175 grams

205 grams of water per gram of stainless steel × 3,175 = 650,875 grams of water

650,875 grams of water × (1 gallon / 3,785.4 grams) = 171.9 gallons for flatware set for 8 people, use 172 ÷ 2 = 86 for 4

Tables

5-foot dining room table: 59 × 35.4 × 29.1 inches (assume 0.5-inch thickness) = 1,509 cubic inches of wood = 10.5 board-feet × 5.4 = 56.7 gallons of water

Chapter 9. Health and Beauty

Introduction

Americans spend more on beauty than on education:
http://www.economist.com/displaystory.cfm?story_id=1795852

Breakdown of US expenditures on beauty products:
http://www.beautyexpo.com.my/article.cfm?id=271

Lipstick:

Estimate of females over the age of 15 in 2009 = 125,701,017:
http://en.wikipedia.org/wiki/Demographics_of_the_United_States#Age_structure

0.72 liter (3 cups) of water per finished cosmetic product:

http://www.environmentalleader.com/2009/04/23/loreal-to-reduce-ghg-emissions-water-consumption-and-waste-by-50

3 cups × 125,701,017 sticks of lipstick = 2.388 million gallons

Blush container:

http://www.victorie-inc.us/mineral_makeup_price_comparison.html

0.6 gram of blush (mineral makeup) weighs 0.02 ounce and is sold in a 3-gram container.

Assume 3 liters of water used for every 1 liter of plastic:

http://www.triplepundit.com/pages/askpablo-glass.php

An empty polyethylene terephthalate bottle weighs 28 to 32 grams (use 30 grams), so the plastic container for blush is $\frac{1}{10}$th the amount of plastic of a bottle. So $\frac{1}{10}$th the water is 0.3 liter = 10 ounces.

Assume half of all women between 20 and 64 have blush:

90,813,578 × 0.5 × 10 ounces = 454,067,890 ounces of water to make containers = 28,379,243.125 pounds = 3,400,640.501 gallons

Products

Antiperspirants

The bulk of the antiperspirant stick formulation consists of waxy or fatty materials, namely castor oil.

Assume a 2.8-ounce stick of deodorant is 50 percent castor oil

2,516 gallons per pound × ($\frac{1}{16}$) × 1.4 ounces = 220 gallons:

http://www.madehow.com/Volume-5/Antiperspirant-Deodorant-Stick.html

Possible health risks posed by antiperspirants:

http://www.controlyourimpact.com/articles/deodorants-antiperspirants-and-your-health

Less-processed food and less meat may lessen odor:

http://www.copperwiki.org/index.php/Deodorant

Cosmetics

Castor beans = 10,500 liters per kilogram, then 1,258.177 gallons per pound. At 50 percent oil yield from the seeds, this is 2,516 gallons per pound

Average virtual water content of primary crops (cubic meters per ton):

http://www.waterfootprint.org/?page=files/NationalStatistics, Appendix XV (global average)

Castor oil yield = 50 percent:

http://www.journeytoforever.org/biodiesel_yield.html

Moisturizers

Water content is typically about 60 percent of total. For 6 ounces, this is 3.6 ounces.

Coconut virtual water content = 2,545 liters per kilogram, then 305 gallons per pound. At 40 percent oil recovery, this is 762.4 gallons per pound of oil.

Coconut oil recovery is 0.4 kilogram of oil per kilogram of coconut flesh:

Rosillo-Calle, Frank, de Groot, Peter, Hemstock, Sarah L., and Woods, Jeremy, eds. 2007. *The biomass assessment handbook: Bioenergy for a sustainable environment.* Sterling, VA: Earthscan.

Cocoa beans require 27,218 liters per kilogram, so 3,261 gallons per pound.

1 kilogram of cocoa beans contains 0.871 kilograms nib, which is 55 percent cocoa butter, so the water footprint of cocoa butter is 6,808.13 gallons per pound

Hui, Y.H., ed. 2005. *Handbook of food science, technology, and engineering: Volume 4.* Boca Raton, FL: Taylor and Francis.

Perfumes

Water content of different fragrance grades:

http://www.enotes.com/how-products-encyclopedia/perfume

Essential oils:

Kourik, Robert. 1998. *The lavender garden: Beautiful varieties to grow and gather.* San Francisco: Chronicle Books.

190 petals = ½ ounce and the typical rose has about 30 petals:

http://www.petalgarden.com/rose-petals-info.htm

300 pounds of petals = 4,800 ounces = 1.824 million petals = 60,800 roses

1 rose has a virtual water content of 10 liters, so 2.64 gallons:

http://digitaluv.com/andreas/blog/?p=345

60,800 roses × 2.64 gallons = 160,616.6 gallons of water per ounce of rose oil.

Rock hyrax:

http://en.wikipedia.org/wiki/Perfume

Shampoos

Shampoos contain 70 to 80 percent water

How it's made:

http://www.enotes.com/how-products-encyclopedia/shampoo

Jojoba seeds: 12,344 liters per kilogram, then 1,479.1 gallons per pound

Soaps

For a 3.1-ounce bar of soap made of 35 percent tallow, 30 percent lard, 30 percent coconut, and 5 percent castor oil:

(1.085 ounces of beef tallow × 1,122.18 gallons per pound) / 16 = 76 gallons per bar of soap

(0.93 ounce of lard [assume pig] × 614) / 16 = 35.7 gallons per bar of soap

(0.93 ounce of coconut × 762.4) / 16 = 44.3 gallons per bar of soap

(0.155 ounce of castor oil × 2,516.35) / 16 = 24.38 gallons per bar of soap

3.1-ounce bar of soap's total virtual water content = 76 + 35.7 + 44.3 + 24.38 = 180.4 gallons per bar

Ingredients in body wash:

http://www.epinions.com/content_2232262788

Bar soap is more economical than body wash (if hard milled):

http://www.associatedcontent.com/article/990424/bar_soap_or_body_wash_which_is_better.html?cat=69

Toothpaste

http://www.enotes.com/how-products-encyclopedia/toothpaste

50 percent water: 4.4-ounce (124 gram) tube: 0.032757334354 gallon × 0.5 = 0.016 gallon = 2.048 fluid ounces of water per 4.4-ounce tube:

http://www.onlineconversion.com/waterweight.htm

Blob the size of a pea:

http://planetgreen.discovery.com/food-health/green-toothpaste.html

Chapter 10. School and Office Products

Introduction

The average office worker uses 10,000 sheets (100 pounds) per year:

http://eetd.lbl.gov/paper/html/concept.htm

1 ton of virgin uncoated paper (90 percent of US printing and writing paper) requires 19,075 gallons of water (9.5375 gallons per pound) + 9.216 gallons per pound to grow the trees, so 18.7535:

http://www.techsoup.org/learningcenter/techplan/page5675.cfm

100 percent recycled paper uses 5.1625 gallons per pound:

http://www.edf.org/papercalculator

If the average office worker uses 100 pounds of copy paper per year and 33 percent of this is from recycled materials, the water footprint is 100 pounds × 0.67 × 18.7535 + 100 × 0.33 × 5.1625 gallons per pound = 1426.85 gallons per office worker:

http://www.epa.gov/waste/conserve/materials/paper/faqs.htm

4 millions tons of copy paper used a year × 33 percent of that being made from trees × 19,075 gallons per ton of virgin = 25,179,000,000 gallons of water for making copy paper from trees (as opposed to wood chips and sawdust) each year:

http://www.epa.gov/waste/conserve/materials/paper/faqs.htm#where

2 billion books published each year:

http://www.epa.gov/waste/conserve/materials/paper/faqs.htm#use

Elementary and high schools make up about 15 percent of the market and college textbooks about 14 percent, so about 30 percent of book publishing is textbooks for schools:

Greco, Albert N., Rodríguez, Clara E., and Wharton, Robert M. 2007. *The culture and commerce of publishing in the 21st century.* Stanford, CA: Stanford Business Books.

Products

Books

http://www.waterfootprint.org/?page=files/productgallery&product=paper

Assume the average textbook weighs 3 pounds:

http://www2.cde.ca.gov/be/ag/ag/may04item21.pdf

See paper entry below for details:

Assume 2/3 is from virgin paper: 18.735 × 2 = 37.47

One-third is from recycled paper: 5.1625 × 1 = 5.1625

37.5 + 5.2 = 42.7

Computers

Microchip: One 2-gram microchip uses 32 liters of water, so 8.4 gallons:

http://www.waterfootprint.org/Reports/Hoekstra_and_Chapagain_2006.pdf

Hewlett-Packard Designing for Environment:

http://www.hp.com/hpinfo/globalcitizenship/environment/productdesign/
materialuse.html

Using price of $500 brings: 500 × 80 = 40,000 liters (at 80 liters per US$) =
10,556 gallons

Price range: $500 to $2,000 depending on make, model, customizing:

http://www.pcworld.com/shopping/browse/category.html?id=10006

$2,000 × 80 = 160,000 liters = 42,267 gallons

Erasers

Assume one pencil eraser weighs 0.1 ounce.

55 gallons per pound of synthetic rubber:

http://www.watersmartsystems.com/WaterTrivia.pdf

7,636 cubic meters per ton = 915 gallons per pound. Allow 20 grams per eraser, so
40.3:

http://www.ias.ac.in/currsci/oct252007/1093.pdf

http://en.wikipedia.org/wiki/Eraser

55 gallons per pound of synthetic rubber, so 2.425 gallons per eraser:

http://www.nypirg.org/ENVIRO/water/facts.html

Paper

1 ton of virgin uncoated paper (90 percent of US printing and writing paper)
requires 19,075 gallons of water (9.5375 gallons per pound) + 9.216 gallons per
pound to grow the trees, so 18.7535 gallons per pound, or 0.18735 gallons per
sheet, or 3 cups per sheet:

http://www.techsoup.org/learningcenter/techplan/page5675.cfm

100 percent recycled paper uses only 5.1625 gallons per pound

http://www.edf.org/papercalculator

http://www.epa.gov/wastes/conserve/materials/paper/index.htm

http://www.waterfootprint.org/?page=files/productgallery&product=paper

Pencils

14 billion pencils made every year:

http://www.newdream.org/marketplace/pencils.php

Most pencils made from California incense cedar:

http://www.greenseal.org/resources/reports/CGR_officesupplies.pdf

These trees have an average height of 105 feet and a diameter of 4 feet:

http://www.nearctica.com/trees/conifer/cupress/Ldecur.htm

This equals 330 cubic feet. There are 12 board-feet in a cubic foot, so one tree
    contains 3,960 board-feet:
http://www.na.fs.fed.us/spfo/pubs/uf/lab_exercises/calc_board_footage.htm
Assume 5.4 gallons per board-foot:
http://www.nypirg.org/ENVIRO/water/facts.html
3,960 × 5.4 gallons = 21,384 gallons per tree
One California incense cedar makes 172,000 pencils:
http://www.greenseal.org/resources/reports/CGR_officesupplies.pdf
21,384 gallons per 172,000 pencils = 0.124 gallon of water per pencil for wood.
55 gallons per pound of synthetic rubber:
http://www.nypirg.org/ENVIRO/water/facts.html
Allow for eraser to weigh 0.1 ounce, so 0.343
Total for pencil + eraser = 0.467 gallons per pencil
14 billion pencils × 0.467 gallons = 6.538 billion gallons total
6.538 billion gallons is 874,003,472 cubic feet.
Chicago's Willis (formerly Sears) Tower occupies 53.4 million cubic feet of space.
874 million cubic feet ÷ 53.4 million cubic feet = filled 16.367 times
500 million gallons per year for Albuquerque school district:
http://www.cabq.gov/aes/s4p2sch.html

Pens

Pen weighs 13 grams:
http://www.nibs.com/OtherPensAndWritingImpPage.htm
294.2 grams of water per gram of polyethylene terephthalate plastic:
http://www.triplepundit.com/pages/askpablo-glass.php
3,822 grams of water + 1 milliliter (negligible in terms of gallons) of ink capacity:
http://en.wikipedia.org/wiki/Fountain_pen
Divided by 3,785.4 grams per gallon = approximately 1 gallon
US gallon = 3.78541178 liters, breaking that into milliliters (3,785.41178). A milliliter
    of pure water (hydrogen oxide, no minerals or anything, which water almost
    always has) weighs exactly a gram. Therefore, 3,785.31178 milliliters of water
    weighs 3,785.41178 grams.

Printers

Ink cartridges:
http://www.castleink.com/_a-recycle-ink-cartridges.html
Each year, millions of empty toner and inkjet cartridges are thrown into the trash,
    ending up in our planet's landfills or incinerators.
Ink cartridges are constructed out of plastic, petroleum-based products and take
    about 1,000 years to decompose. According to recent estimates, 20 to 40 percent
    of ink cartridges are recycled, meaning 60 to 80 percent end up in landfills.
Waterless printing:
http://www.waterless.org/NwhatIs/whatIs.htm
$450 is average color laser printer price:
http://laserprinterprice.com

80 liters per US$1 = 36,000 liters or 9,510.2 gallons:
http://www.waterfootprint.org/?page=files/productgallery&product=industrial
Tape
http://en.wikipedia.org/wiki/Scotch_Tape
http://solutions.3m.com/wps/portal/3M/en_US/Manufacturing/Industry/Product-Catalog/Tapes
http://www.answers.com/topic/cellophane
http://www.shop3m.com/70016031984.html
Allow 1 cup per 1,296-inch-long, 3/4-inch-wide roll of clear tape per process wash (6 process washes to make and form)

Chapter 11. Luxury

Introduction

Diamond engagement ring sales were $6.2 billion in 2006, at an average cost of $3,200 per engagement ring:
http://www.novori.com/news/diamond-sales.html
$6.2 billion at $3,200 gives an average of 1.94 million rings sold
De Beers used 41.4 million cubic meters of potable water + 20.9 million cubic meters of reused water + 59.7 million cubic meters in 2007 for mining operations for a total of 122 million cubic meters:
http://www.debeersgroup.com/ImageVault/Images/id_1739/ImageVaultHandler.aspx
In 2007 De Beers produced 51.1 million carats:
http://www.jckonline.com/article/281471-De_Beers_Diamond_Sales_Down_3_in_2007.php
122 million cubic meters ÷ 51.1 million carats = 2.387 cubic meters per carat = 630.7 gallons per carat
250,000 cubic meters per ton = 250,000 liters per kilogram:
http://www.infomine.com/publications/docs/InternationalMining/Chadwick2007o.pdf
Assume a gold ring contains 0.4 ounce gold = 0.0113398093 kilograms
One 0.4-ounce ring contains 748.91 gallons of water.
A 1-carat diamond ring therefore contains 630.7 gallons per diamond + 748.91 gallons per gold ring = 1,379.6 gallons per 1-carat diamond ring
1,379.61 × 1.94 million diamond engagement rings = 2.676 billion gallons
"There are more than 2,000 yachts in the world that measure 100 feet, but if you want to make a real impression, your yacht needs to measure at least 200 feet":
http://www.msnbc.msn.com/id/13345720
Lamborghini Murcielago gets a paltry 9 miles to the gallon:
http://www.thedailygreen.com/living-green/blogs/cars-transportation/exotic-supercars-gas-mileage-460221

Price tag of $382,400 × 100 liters per $1, so 10,101,939.3 gallons:

http://usnews.rankingsandreviews.com/cars-trucks/Lamborghini_Murcielago

Most subsonic aircraft water vapor emissions are removed from the atmosphere
through precipitation within 1 to 2 weeks:

http://www.co2offsetresearch.org/aviation/DirectEmissions.html

1 kilogram of kerosene releases 1.25 kilograms of water (1.25 liters or 0.330215065
gallons):

http://www.mpimet.mpg.de/en/presse/faqs/welche-rolle-spielen-kondensstreifen-
fuer-unser-klima.html

Kerosene density: 817.15 kilograms per cubic meter, so 0.81715 kilogram per liter or
1.223765 liters per kilogram or 0.323284651 gallon per kilogram:

http://www.simetric.co.uk/si_liquids.htm

0.323284651 gallon of kerosene yields 0.330215065 gallon of water vapor

1.0214375 gallons of water vapor per gallon of kerosene

Assume 5 gallons per mile flown:

http://www.howstuffworks.com/question192.htm

5 gallons × 1.0214375 = 5.107 gallons of water vapor per mile flown

Wine

120 liters per 125-milliliter glass, so 600 liters per 750-milliliter bottle:

http://www.waterfootprint.org/?page=files/productgallery&product=wine

600 liters = 158.5 gallons per bottle

A 2,000-bottle collection = 158.5 gallons per bottle × 2,000 bottles = 317,006.5
gallons

Cars

39,000 gallons to make a car:

http://www.nypirg.org/ENVIRO/water/facts.html

Conversion from dollars to liters of industrial products made in different countries:

http://www.waterfootprint.org/?page=files/productgallery&product=industrial

Jets

New private long-range jets cost $40 million to $70 million:

http://www.airtravelgenius.com/ar/corporate_jets_sale.htm

Assume 100 liters for every US$1, so 26.4172052 gallons:

http://www.waterfootprint.org/?page=files/productgallery&product=industrial

Total water footprint: 1.057 billion to 1.85 billion gallons

Jewelry

Use 630.7 gallons per carat (see calculation for diamond rings above)

Assume each earring contains 0.2 ounce of gold. Use 748.91 gallons of water per
0.4 ounce gold (see calculation for diamond rings above)

630.7 × 2 + 748.91 = 2010.31

Recycled jewelry:

http://www.brilliantearth.com/recycled-gold-jewelry

Televisions

$185 to $3,100 for a 23-inch LCD to a 55-inch LED:
http://www.pcworld.com/shopping/browse/category.html?id=10129
Conversion from dollars to liters of industrial products made in different countries:
http://www.waterfootprint.org/?page=files/productgallery&product=industrial
185 × 80 = 14,800 liters = 3,909.7 gallons
3,100 × 80 = 248,000 liters = 65,541.7 gallons
Raw materials:
http://www.enotes.com/how-products-encyclopedia/television
Metals used in televisions: steel, copper, lead, tin, aluminum, gold:
http://wiki.answers.com/Q/What_metals_are_used_in_a_television_set

Watches

Conversion from dollars to liters of industrial products made in different countries:
http://www.waterfootprint.org/?page=files/productgallery&product=industrial
$87 × 80 = 6,960 liters = 1 gallon or 3.785 liters = 1,838.8 gallons

Yachts

http://www.yachtworld.com/core/listing/pl_boat_detail.jsp?hosturl=hmyyachtsale
    s&checked_boats=2066266&featuredon=yw-en&slot=3&slim=broker
Conversion from dollars to liters of industrial products made in different countries:
http://www.waterfootprint.org/?page=files/productgallery&product=industrial
At 80 liters per US$1, $13 million takes 1.04 billion liters = 274.7 million gallons

Chapter 12. Pets

Introduction

Pet ownership statistics (2009):
http://www.americanpetproducts.org/press_industrytrends.asp
It costs about $730 per year to own a dog:
http://www.familyresource.com/lifestyles/pets/how-much-does-it-cost-to-own-a-
    pet
A dog should drink 1 ounce per pound of body weight per day. So if the average dog
    weighs 30 pounds, this is 30 ounces per dog:
http://www.ehow.com/facts_5135015_much-should-dog-drink-day.html
30 ounces × 77.5 million = 18.16 million gallons
38 percent of dog owners don't pick up after their pets. Pet waste contributes 20
    to 30 percent of water pollution:
http://www.heraldtribune.com/article/20090730/COLUMNIST/
    907301017?Title=To-protect-the-bay-pick-up-after-your-pet
Escherichia coli on California beaches; 5,000 pounds of solid waste per day at the
    Four Mile Run:
http://www.jupiterionizers.com/catalog/article_info.php?tpath=16&articles_id=37

3.6 billion pounds of dog waste per year would fill 800 football fields 1 foot high:
http://www.cmbarkpark.org/NewsLetters/Issue35.pdf
$27.6 billion spent for pet products and supplies and $14.3 billion for vet care and services
Quotes from the American Pet Products Association:
http://www.americanpetproducts.org/press_industrytrends.asp

Beds

Assume size: 27 × 21 × 5 inches = 1,614 square inches of surface area for 100 percent organic cotton/canvas cover:
http://www.discount-pet-mall.com/pet-products/barbados-comfort-dog-bed.html
1,614 square inches of cotton canvas, or 1.25 square yards. Round up to 1.5 square yard since fabric must be sewn together.
Assume canvas is 10 ounces per square yard:
http://canvas-tarps.com
1.5 square yard = 15 ounces = roughly 1 pound
Assume cotton is 1,300.12 gallons per pound:
http://www.waterfootprint.org/Reports/Report18.pdf
Assume 12 pounds of polyester insert.
Assume virgin polyester uses 29.5 gallons per pound and recycled polyester uses 26 percent less water than virgin polyester:
http://asint2.ki.com/PROD/PKB/cstmrpkb.nsf/a99d8ef084f3a149862568e700715882/a3c970a25676fca3862573400050645f/$FILE/terratext_lifecycle.pdf
29.5 × 12 + 1,300 = 1,654 gallons

Bowls

5.3 grams of water per gram of ceramic:
http://www.triplepundit.com/pages/ask-pablo-the-c.php
Assume a medium ceramic pet bowl weighs 4 pounds:
http://www.kennelvet.com/feeders-waterers-for-pets-luxury-ceramic-dog-bowls-c-1682_1953.html
4 pounds = 1,814 grams
1,814 × 5.3 = 9,616 grams of water = 9,616 milliliters of water = 9.616 liters = 2.54 gallons

Collars

16,600 liters for 1 kilogram of leather, so 1,989.12 gallons per pound:
http://www.waterfootprint.org/?page=files/productgallery&product=leather
Assume a leather collar weighs 0.125 pound, so 248.6 gallons

Food

Canned foods typically contain 75 to 78 percent moisture, whereas dry foods
contain only 10 to 12 percent water:
http://www.fda.gov/AnimalVeterinary/ResourcesforYou/ucm047113.htm

Assume that a vegetarian pet food would have a water footprint of no greater than
that of cereal (which is extruded in the same way pet food is manufactured):
460 gallons per pound

Assume that a meat diet would have a water footprint no greater than that of pure
beef: 1,560 gallons per pound

Hoekstra, Arjen Y., and Chapagain, Ashok K. (2008) *Globalization of water: Sharing
the planet's freshwater resources*. Malden, MA: Blackwell Publishing. US data
from http://www.waterfootprint.org/?page=files/NationalStatistics, Appendix
XVI, p. 16.

Leashes

Assume nylon consumes the same water as other "poly" manufacturing:
294.2 milliliters of water per gram of plastic, so 35.25 gallons per pound:
http://www.triplepundit.com/2007/03/askpablo-glass-vs-pet-bottles

Average 6-foot leash weighs ½ pound

$35.25 \times 0.5 = 17.6$ gallons

Toys

Assume a hemp rope weighs 0.75 pound. Hemp = 333.72 gallons per pound
250.3 gallons for 0.75 pound

Treats

http://www.bullysticks4dogs.com/bully_sticks_made.htm

A bully stick weighs 0.25 pound. Using 1,580 gallons for unprocessed beef per
pound = 395.2 gallons:
http://www.waterfootprint.org/?page=files/NationalStatistics

Potato = 12.7 gallons per pound:
http://www.waterfootprint.org/?page=files/NationalStatistics

Poop Bags

It takes 58 gallons to produce 1,500 plastic bags. This is 5 ounces per bag:
http://www.savetheplasticbag.com/ReadContent486.aspx

300 million pounds of plastic from 14.6 million bushels of ground corn:
http://www.popularmechanics.com/science/research/1281896.html

0.048667 bushels per pound of plastic

45 pounds per bushel of ground corn:
http://www.sizes.com/units/bushel_US_as_mass.htm

0.048667 × 45 = 2.19 pounds of corn per pound of plastic

58.6 gallons per pound of corn

58.6 × 2.19 = 128.3 gallons per pound of plastic made from corn

200 biobags weigh 6.9 ounces (0.43125 pound)

One biobag weighs 0.0021563 pound

Production of one biobag from corn uses 0.0021563 pound × 128.3 gallons per pound = 0.277 gallons, so 35 ounces

Chapter 13. Building Materials and Appliances

Introduction

The average person moves every 5 to 7 years:
http://homebuying.about.com/od/sellingahouse/qt/0207WhyMove.htm

1.6 million gallons per home × 1.5 million new homes a year = 240 billion gallons

240 billion gallons = 3.21010 cubic feet = 3.21010 square feet × 1 foot depth

3.2 × 1,010 square feet = 1,150.82 square miles

Rhode Island is 1,545 square miles:
http://en.wikipedia.org/wiki/Rhode_Island

Tool Belt

Leather tool belt weighs 5.6 pounds:
http://www.bestbelt.com/product/toolbelts/5089-toolbelt.html

Cow leather is 1,989.12 gallons per pound

5.6 × 1,989.12 = 11,139 gallons

Hammer

7 to 20 ounces for the steel head and 12 to 13 inches for the handle (can be of wood):
http://www.madehow.com/Volume-4/Hammer.html

Handle: assume 1 inch in diameter

12 inches × pi (1)2= 37.7 cubic inches = 0.26 board-feet × 5.4 gallons per board-foot = 1.404 gallons for handle

Head: assume medium-size head of 14 ounces = 14 gallons

Steel in a 30-pound bicycle takes 480 gallons of water = 16 gallons per pound = 1 gallon per ounce:
http://www.nypirg.org/ENVIRO/water/facts.html

14 gallons + 1.404 gallons = 15.4 gallons for a hammer

Cabinetry

Wood

Cabinetry in the average kitchen is 12 feet of wall cabinets (34 inches high) and 13 feet of floor cabinets (36 inches high):
http://www.demesne.info/Improve-Your-Home/Kitchens/Kitchen-cabinets.htm

If each cabinet is 1 inch thick, this is 73 board-feet:

http://northernloghome.com/lf-2-bf.htm

At 5.4 gallons of water per board-foot, 395 gallons of water would be consumed:

http://www.chnep.org/MoreInfo/water_conservation_facts.htm

Medium-Density Fiberboard (MDF)

MDF weighs 4.33 pounds per board-foot; 73 board-feet weighs 317.6 pounds:

http://www.mcquesten.com/resources/product_weights.html

MDF manufacturing can consume up to 10,000 gallons of water per ton of product.
So 317.6 pounds ÷ 2,000 × 10,000 = 1,588 gallons:

http://www.fpl.fs.fed.us/documnts/fplrn/fplrn077.pdf

Dangers of MDF:

http://www.greenseal.org/resources/reports/CGR_particleboard.pdf

Carpeting

Polyester is the most widely used fiber for residential carpets:

http://www.simplyfloored.com/about_flooring/carpet.asp

50 million pounds of recycled polyester uses 74 million gallons less water than
virgin polyester would require. This 74 million gallons is 73 percent of the
water used for virgin polyester. So 50 million pounds of virgin polyester uses
74 million gallons ÷ 0.27 = 274 million gallons.

274 million gallons per 50 pounds of virgin = 1 pound of virgin polyester consumes
5.5 gallons of water:

http://asint2.ki.com/PROD/PKB/cstmrpkb.nsf/a99d8ef084f3a149862568e700715
882/a3c970a25676fca3862573400050645f/$FILE/terratext_lifecycle.pdf

Polyester is made from plastic, and it takes 24 gallons of water to produce 1 pound
of plastic:

http://www.chnep.org/MoreInfo/water_conservation_facts.htm

Assume ½ pound of carpet per square foot. Each square foot requires 14.75 gallons
of water. For 1,000 square feet of carpeting, this is 14,750 gallons of water.
Forty 2-liter bottles per square yard of carpet, or 4.44 per square foot. So
1,000 square feet of recycled carpet would save 4,440 two-liter bottles:

http://www.greenflooringconcepts.com/products.htm#crc

50 million pounds of recycled = 200 million gallons (274 × 0.73)

200 million gallons ÷ 50 million pounds = 4 gallons per pound

Assume ½ pound per square foot of carpet, so for recycled:

2 gallons per square foot × 1,000 square feet = 2,000 gallons, which is 86.5
percent of 14,750

Clothes Dryers

Components of clothes dryer:

http://home.howstuffworks.com/dryer.htm

80 liters per $1 (waterfootprint.org) × 3.785 liters per gallon = $0.047 per gallon
= 21.136 gallons per $1 × 800 = 16,909 gallons

Granite Countertops

Granite production consumes 9,810 gallons of water per ton, which is 9,810 gallons per 2,000 pounds, so 4.9 gallons per pound:
http://isse.utk.edu/ccp/projects/naturalstone/pdfs/MS_Granite.pdf
The average kitchen has 50 to 60 square feet of counter space:
http://ezinearticles.com/?How-Much-Will-My-Laminate-Countertop-Cost?&id=850275
The density of granite is 160 pounds per cubic foot. Assuming the average counter is 1 inch thick, the weight of 1 square foot of granite countertop = 160 ÷ 12 = 13.33 pounds per square foot:
http://isse.utk.edu/ccp/projects/naturalstone/pdfs/MS_Granite.pdf
13.33 pounds per square foot × 60 square feet = 800 pounds. The water embedded in 800 pounds of granite = 800 × 4.9 = 3,920 gallons.

Gutters

Aluminum

The average American house requires 120 feet of rain gutter:
http://www.etup.org/articles/article-493.html
Aluminum gutter material is sold by the coil. A 990-foot coil weighs 350 pounds. So this is 350 ÷ 990 = 0.3535 pounds per linear foot:
http://kroybp.com/WebDocument.nsf/LookupSpecs/AluminumRainware/$file/AlumRainware.pdf
120 feet of rain gutter weighs: 120 × 0.3535 = 42 pounds.
Based on the average water consumption for primary smelting from these two sources, we assume aluminum production from virgin materials is 72 gallons per pound:
http://www.alcoa.com/canada/en/pdf/alcoa_2007_eng.pdf
http://www.epha.eg.net/pdf/n3-4-2006/4-Manal%20Hosny-Life%20cycle.pdf
Based on the average water consumption for secondary smelting from these two sources, we assume aluminum production from recycled materials is 5.5 gallons per pound.
We'll take the average for secondary smelting and call it 5.5 gallons per pound:
http://www.veoliawater.com/services/industrial-customers/industrial-sectors/metals/#c11938456001
Smelting and processing into a final product uses 3 gallons per pound:
http://www.epha.eg.net/pdf/n3-4-2006/4-Manal%20Hosny-Life%20cycle.pdf
Aluminum containing 50 percent recycled content would use 39 gallons per pound.
42 pounds × 39 gallons per pound = 1,638 gallons

Copper

Assume 1 pound of copper per linear foot:

http://www.guttersupply.com/p-Copper-Clad-Stainless-Steel-Coil.gstml

A house with 120 feet of copper gutters would require 120 pounds of copper. At 38.5 gallons of water per pound of copper (see copper piping for conversions), this is 38.5×120 pounds = 4,620 gallons of water.

Hardwood Flooring

1,000 board-feet of flooring will cover 750 square feet. So 1,000 square feet of flooring requires 1,333 board-feet:

http://www.harleyvillebuilders.com/FloorEstimating.html

At 5.4 gallons per board-foot, this is 7,200 gallons:

http://www.chnep.org/MoreInfo/water_conservation_facts.htm

Insulation

3.907 liters per square meter of fiberglass = 0.09589 gallon per square foot

0.822 liters per square meter for paper = 0.02 gallon per square foot

$0.09589 \times 2,500 = 239.7$ and $0.02 \times 2,500 = 50$:

http://www.springerlink.com/content/h0j2j8wn5604j0g9

http://www.cumberlandcountyhomebuilders.com/nahb.php

Linoleum Flooring

Linoleum manufacturing uses 157.5 liters per square meter. At 0.264 gallon per liter and 10.76 square feet per square meter, this equates to about 3.866 gallons per square foot, or 3,865 gallons per 1,000 square feet:

http://www.inies.fr/indic_prod.asp?id_prod=95&mode=Fam

Vinyl production requires 30 gallons of water per pound:

http://www.buildinggreen.com/auth/article.cfm?fileName=030101b.xml

Assume vinyl flooring weighs 0.44 pound per square foot. So 1,000 square feet would require 13,333 gallons of water:

http://www.fastfloors.com/catalog/productline.asp?productlineid=20338&product id=142629&REF=ZLS3998017

Ovens and Stoves

Using 80 liters per US$1 = 21.136 gallons per 1×650 = 13,738.4 gallons

Piping

0.27 percent copper in 1 ton of ore:

http://www.cat.com/cda/files/561955/7

500 gallons per ton of ore = 500 gallons per 2,000 pounds of ore

2,000 pounds of ore × 0.0027 = 5.4 pounds of copper per ton of ore:
http://www.epa.gov/osw/nonhaz/industrial/special/mining/techdocs/copper.pdf

Refrigerators

Components of a refrigerator:
http://css.snre.umich.edu/css_doc/CSS04-13.pdf
1,200 × 21.136 = 25,363.2 gallons

Roofing

1 barrel of oil is 42 gallons and yields 11 gallons of asphalt and other products:
http://wiki.answers.com/Q/How_much_gasoline_can_be_made_from_one_barrel_
of_crude_oil
Asphalt weighs 145 pounds per cubic foot:
http://wiki.answers.com/Q/How_much_does_1_cubic_yard_of_asphalt_weigh
1,851 gallons of water per barrel of oil × 1 barrel per 11 gallons of asphalt × 1 gallon /
0.134 cubic foot × 1 cubic foot ÷ 145 pounds = 8.6 gallons per pound of
asphalt
2.3 pounds of asphalt per square foot × 2,000 square feet × 8.6 gallons per square
foot = 39,844.6 gallons
The average 1,700-square-foot home requires 1,992 square feet of roofing:
http://www.cumberlandcountyhomebuilders.com/nahb.php
Cement production uses 1,360 gallons of water per ton. At 2,000 pounds per ton,
this is 0.68 gallons per pound. Lightweight concrete tile weighs 5 pounds per
square foot. A 2,000-square-foot roof requires 10,000 pounds of concrete tiles
and 6,800 gallons of water:
http://www.chnep.org/MoreInfo/water_conservation_facts.htm
http://www.hometips.com/masonry_roof.html
Clay production uses 2,000 gallons per ton of finished product: 2,000 pounds
per ton. Therefore, 1 gallon per pound of clay. Clay tile roofs weigh 788
pounds per square. One square = 100 square feet. 2,000 square feet of
roofing requires 15,760 pounds of clay tile, which consumes 15,760 gallons of
water:
http://www1.eere.energy.gov/industry/mining/pdfs/water_use_mining.pdf
http://products.construction.com/Manufacturer/MCA-Clay-Roof-Tile-NST2577/
products/One-Piece—S—Mission-Tile-NST8179-P

Stone-Tile Flooring

Limestone is 19,600 gallons per ton:
http://isse.utk.edu/ccp/projects/naturalstone/pdfs/MFS_Limestone.pdf
Assume 6 pounds per square feet × 1,000 square feet = 6,000 pounds = 3 tons
19,600 × 3 = 58,800 gallons

Tile Countertops

The density of ceramic tile is 3.25 pounds per square foot. So 60 square feet requires 190 pounds of ceramic tile:

http://www.eurotile.ca/tilespec.asp?pid=1215&dm=

5.3 pounds of water per pound of ceramic:

http://www.triplepundit.com/2006/09/ask-pablo-the-coffee-mug-debacle

190 pounds of ceramic requires 5.3 × 190 = 1,007 pounds of water. Using the conversion factor of 8.33 pounds per gallon, this is 120 gallons per tile counter.

Washing Machines

Raw materials in a washing machine:

http://www.enotes.com/how-products-encyclopedia/washing-machine

Use global average for industrial products: 80 liters per US$1:

http://www.waterfootprint.org/?page=files/productgallery&product=industrial

Assume a washer costs $500 = 10,565 gallons:

http://www.nextag.com/clothes-washer/shop-html

The average top-loader uses 40 gallons per wash. The average front-loader uses 14.7 gallons per wash. Using a figure of 392 loads of laundry per year, the top-loader will use 40 × 392 × 10 = 156,800 gallons and the front-loader will use 14.7 × 392 × 10 = 57,624 gallons:

http://michaelbluejay.com/electricity/laundry.html

Windows

Aluminum Frames

An average aluminum window frame weighs 41 pounds:

http://dsp-psd.pwgsc.gc.ca/Collection/NH18-22-98-120E.pdf

An average house with 12 windows would need 492 pounds of aluminum.

At 39 gallons per pound, this would consume 19,190 gallons of water.

(See Aluminum Gutters for rationale behind this conversion factor.)

Vinyl Frames

In 1992, 280 million pounds of polyvinyl chloride (PVC) were used to create 8.6 million vinyl windows. This equates to 32.5 pounds of vinyl per window:

http://www.buildinggreen.com/auth/article.cfm?fileName=030101b.xml

Vinyl production requires 30 gallons of water per pound:

http://www.buildinggreen.com/auth/article.cfm?fileName=030101b.xml

For 12 windows in an average house, the total amount of vinyl used would be 390 pounds, which would consume 11,700 gallons of water.

Wood Frames

A typical window is 3 feet wide by 5 feet tall and has a perimeter of 16 feet:

http://www.monolithic.com/stories/windows-doors-and-openings

A window frame that's 2 by 4 feet and 16 linear feet requires 10.67 board-feet:
http://www.csgnetwork.com/boardft2linftconv.html?feet=16&width=4&depth=
    2&qty=1&calcvall=

The average 1,700-square-foot house has 12 windows, so it would require 128
    board-feet:

http://www.cumberlandcountyhomebuilders.com/nahb.php

At 5.4 gallons per board-foot, this is 690 gallons:

http://www.chnep.org/MoreInfo/water_conservation_facts.htm

quick guide

ITEM	VIRTUAL WATER CONTENT
Alfalfa sprouts	14.8 gallons per pound
Almonds	259.2 gallons per cup
Antiperspirant (2.4 ounce, stick)	220 gallons each*
Apple	18.5 gallons each*
Apple juice	349.2 gallons per gallon
Apricot (dried)	20 gallons per serving (6 dried apricots)
Apricot (fresh)	19.8 gallons per serving (3 apricots)
Avocado	42.6 gallons each
Banana	17.5 gallons each
Barley	100 to 200 gallons per pound
Beans	56.2 gallons per pound
Bed (queen size)	2,878.3 gallons for a spring mattress*
Beef (boneless cuts)	1,581 gallons per pound
Beef (cuts with bone)	1,122 gallons per pound
Beer (8 ounces)	19.8 gallons
Blanket	3,824 gallons for wool; 166 gallons for acrylic or other synthetic fabric*
Blueberries	13.8 gallons per cup

* Global average

ITEM	VIRTUAL WATER CONTENT
Books	42.8 gallons each*
Bottle (glass, 12 ounces)	1.1 gallons each
Bottle (plastic, 1 liter)	3 liters each*
Bowl (ceramic)	2.5 gallons each*
Broccoli	27.4 gallons per pound
Butter	3,602.3 gallons per pound*
Cabbage	20.8 gallons per head
Cabinets (kitchen, set)	400 gallons for wood; 1,590 gallons for medium-density fiberboard*
Can (12 ounces, aluminum)	1.1 gallons each*
Candy	7.4 or more gallons per pound
Carpeting (synthetic)	14,750 gallons per 1,000 square feet*
Carrots	6.5 gallons per pound
Cars	39,000 gallons each for the steel alone*
Castor oil	2,516 gallons per pound*
Celery	6.5 gallons per pound
Chair (cow leather)	11,000 or more gallons each*
Cheese	414.2 gallons per pound
Cheese (curd-based; cottage/ricotta)	260.5 gallons per pound
Chest	91.5 gallons per 3-drawer wooden*
Chicken	468.3 gallons per pound
Clothes dryer	16,909 gallons each*
Cocoa oil and butter	6,808 gallons per pound*
Coconut oil	762 gallons per pound*
Coconuts	320.6 gallons each
Coffee	37 gallons per cup (brewed)*

* Global average

ITEM	VIRTUAL WATER CONTENT
Computer	10,556 to 42,267 gallons each, depending on type*
Corn	108.1 gallons per pound*
Cornflakes	47.7 gallons per 18 ounces (2.6 gallons per bowlful)
Cosmetics	Varies by product: 3 cups per lipstick; 0.3 liter per blush*
Cotton	1,300.12 gallons per pound*
Couch (cow leather)	35,600 gallons or more each*
Cow leather	1,989.12 gallons per pound*
Cucumbers	28.4 gallons per pound
Diamond ring	748.9 gallons per 0.4-ounce gold band; 630.7 gallons per carat diamond
Diamond stud earrings (1 carat)	2,010 gallons per pair*
Dinnerware (ceramic)	12.7 gallons per dinner set (dinner plate, salad plate, bowl, and mug)*
Egg (large)	22.8 gallons each
Eraser	40.3 gallons each for rubber; 2.4 gallons each for synthetic*
Flax	375.5 gallons per pound
Garlic	0.21 gallon per clove
Gin	114.5 gallons per liter
Glassware (8 ounces)	17 cups each*
Granite countertop	3,920 gallons per 60 square feet of slab or tile about 1 inch thick*
Granola	65 gallons per cup
Grapefruit	16.4 gallons per pound
Grapes	14.8 gallons per bunch
Gutters (120 linear feet)	1,640 gallons for aluminum; 4,620 gallons for copper*

* Global average

ITEM	VIRTUAL WATER CONTENT
Hammer (steel head, wood handle)	15.4 gallons each
Hardwood flooring	7,200 gallons per 1,000 square feet*
Hemp	300 gallons per pound
Insulation (2,500 square feet)	50 gallons for recycled newspaper; 239.7 gallons for fiberglass*
Jacket (cow leather, hip length)	7,956 gallons each*
Jeans or pants	2,866 gallons each*
Jets	1 billion to 2 billion gallons, depending on size and style*
Jojoba	1,479.1 gallons per pound
Kiwifruit	15.4 gallons each
Lamb	398.8 gallons per pound
Lemon	4.8 gallons each
Lettuce	10.4 gallons per pound
Lime	4.8 gallons each
Linoleum flooring	3,865 gallons per 1,000 square feet*
Mango	81.9 gallons each
Melon	15.3 gallons each
Milk (lowfat or nonfat)	720.1 gallons per gallon
Milk (powdered)	3,294 gallons per gallon reconstituted
Milk (whole or cream)	1,317 gallons per gallon
Moisturizer (6 fluid ounces)	3.6 fluid ounces
Mushrooms	Several gallons per pound (indirectly)
Natural rubber	1,564.7 gallons per pound*
Oats	122.7 gallons per pound
Oats (rolled or flaked)	193.8 gallons per pound

* Global average

ITEM	VIRTUAL WATER CONTENT
Onions	25.6 gallons per pound
Orange	13.2 gallons each*
Orange juice	272.2 gallons per gallon
Oven	13,738 gallons each*
Paper	5.16 gallons per pound recycled; 18.8 gallons per pound virgin (3 cups per sheet virgin)
Pasta	230.5 gallons per pound*
Peach or nectarine	11.2 gallons each
Pear	7.8 gallons each
Peas (fresh)	10.2 gallons per cup
Peas (frozen)	14.5 gallons per cup
Pen	1 gallon each*
Pencil	7.5 cups each*
Pepper (black)	589.7 gallons per pound
Peppers (bell)	18.1 gallons per pound
Perfume (5 ounces)	1 ounce for eau de toilette; 0.5 ounces for cologne*
Pet bed (12 pounds, medium size)	1,654 gallons each
Pet collar (leather)	248.6 gallons each
Pet food	460 gallons per pound of dry vegetarian kibble; 1,580 gallons per pound of meaty canned food
Pet leash (6 feet, nylon)	17.6 gallons each
Pet poop bags	5 ounces each for regular; 35 ounces each for corn-based biodegradable
Pet toys (hemp rope)	250.3 gallons each
Pillow	458.5 gallons each for recycled, synthetic fiber; 14,026.2 gallons each for goose down*
Pineapple	34.5 gallons each

* Global average

ITEM	VIRTUAL WATER CONTENT
Pineapple juice	361.2 gallons per gallon
Piping (copper)	6,930 gallons per 280 feet (enough for a 1,700-square-foot home)*
Pizza (margherita, 10 inches)	312 gallons each*
Plum	14.3 gallons each
Popcorn (3 ounces)	9.5 gallons*
Pork	614.3 to 648 gallons per pound
Pork (cured)	676.3 gallons per pound
Pork (sausage, etc.)	1,176.7 gallons per pound
Potato chips (200 grams)	48.9 gallons*
Potatoes	12.7 gallons per pound
Pots and pans (stainless steel)	56.8 gallons per saucepan*
Printer (color)	9,510.2 gallons each*
Raisins	44 gallons per cup
Raspberries	18.4 gallons per cup
Red wine (4 ounces)	31.7 gallons*
Refrigerator (standard side-by-side)	25,363.2 gallons each*
Rice	200 gallons per pound (96 gallons per cup)
Roofing (2,000 square feet)	6,800 gallons for cement; 15,760 gallons for clay; 39,844.6 gallons for composite (asphalt-based)*
Rug (5 by 9 feet, cotton/wool blend)	9,531 gallons each*
Rum	33 gallons per liter
Running sneakers	1,247 gallons per pair*
Rye	39.8 gallons per pound (8.8 gallons per cup)

* Global average

ITEM	VIRTUAL WATER CONTENT
Shampoo (22 ounces)	17 ounces*
Sheep or lamb leather	731 gallons per pound*
Sheets (cotton, queen size)	6,663 gallons for 400 thread count; 9,000 gallons for 1,000 thread count*
Shirt (men's dress, cotton)	975 gallons each*
Shoes (leather)	2,113 gallons per pair*
Silverware (5-piece place set for 4 people)	86 gallons*
Soap (bar, 3.1 ounces)	180.4 gallons each*
Socks	244 gallons per pair*
Soda (plastic bottle, 2 liters)	132 gallons each
Soybeans	224 gallons per pound
Spinach	12.3 gallons per pound
Squash	40.7 gallons per pound
Steel	16 gallons per pound*
Stone tile flooring (limestone or travertine)	58,800 gallons per 1,000 square feet*
Strawberries	3.6 gallons per cup
Sugar (beet)	71.7 gallons per pound
Sugar (cane)	100.4 gallons per pound (1 gallon per teaspoon)
Suit (cashmere)	1,528 gallons each*
Suit (cotton)	3,900 gallons each*
Suit (wool)	2,282.8 gallons each*
Sweater (wool)	594 gallons each*
T-shirt (cotton)	569 gallons each*
Table (dining room)	56.7 gallons each*

* Global average

ITEM	VIRTUAL WATER CONTENT
Tape (clear, 12.5 yards)	6 cups per roll*
Tea (black)	5.5 gallons per brewed cup
Television	3,900 to 65,500 gallons each, depending on size and type*
Tequila	64.7 gallons per liter
Tile countertop (60 square feet)	120 gallons*
Tomato (2.5 ounces)	1.3 gallons each
Tomato juice	87 gallons per gallon
Tool belt (leather)	11,139 gallons each
Toothpaste (4 ounces)	2 fluid ounces each*
Treats (dog)	395.2 gallons per 12-inch bully stick
Turkey	286.3 gallons per pound
Underwear (cotton)	86 gallons each for ladies' bikinis; 252 gallons each for men's boxers*
Vinyl flooring (1,000 square feet)	13,330 gallons*
Vodka (1 liter)	79.8 gallons
Washing machine	10,565 gallons each*
Watch (quartz)	1,800 gallons or more each*
Wheat bran	100 gallons per pound
Wheat or white flour	101.7 gallons per pound
Whiskey (1 liter)	430 gallons
White wine (4 ounces)	28.5 gallons*
Window frame	691.4 gallons each for wood; 11,700 gallons each for vinyl; 19,190 gallons each for aluminum*
Yacht (120 feet, $13 million)	275 million gallons each*
Yogurt (6 ounces)	36.3 gallons each

* Global average

acknowledgments

Dr. Marah Hardt took the lead in researching and helping to write this book. I owe her much gratitude and praise, and this space is too little to suffice. Charles Sharp also assisted in the research of the many references, links, and resources found in these pages. Nice work. Susan Raihofer helped shape this book from idea to fruition. As always, she makes things happen. (Family-friendly verbiage.) Colin Dickerman needs many thanks and much appreciation for his nod and getting the book going. I appreciate the whole team at and for Rodale seeing things through: Gena Smith, Yelena Gitlin, Chris Krogermeier, Sara Cox, and, by thankful association, Nancy Elgin.

Jeannie Lee, my family and friends are my support. I appreciate you all being there for me (even though I may not say it so much). Ditto for Margaret Riley and Lauren Auslander.

Maude Barlow, John Dickerson, Alexandra Cousteau, and all those who in different ways work tirelessly to get the word out about water—I salute you.

Arjen Hoekstra deserves more than an acknowledgment; he deserves credit for the backbone, much of the data, and the original thinking on water and virtual water issues. I can't thank you enough, sir. I and my research staff leaned on www.waterfootprint. org a lot. I encourage others to visit the site for deeper research.

index

A

Aerators, faucet, 9, 11, 138
Air conditioners, 9, 138
Airports, water-saving ideas for,
36-37, 146-47
Alcoholic beverages
    beer, 64, 160-61
    gin, 64, 161
    rum, 64, 161-62
    tequila, 64-65, 162
    vodka, 65, 162
    whiskey, 65, 162
    wine, 64, 65, 119, 161, 162, 177
Alfalfa sprouts, 51, 153
Almonds, 49, 151-52
Aluminum
    cans, 61, 159
    gutters, 183
    windows, 109, 186
American Pet Products Association
    (APPA), 98-99
Animals, humane treatment of, 55
Antiperspirants, 80-81, 171
APPA, 98-99
Apple juice, 62, 160
Apples, 45
Appliances
    clothes dryers, 105-6, 182-83
    ovens and stoves, 107, 184
    references, 137
    refrigerators, 107, 185
    washing machines, 108, 186
Apricots, 45-46, 149
Avocados, 46, 149

B

Bacon, 114
Bananas, 46, 149-50
Barley, 57, 64, 156
Baseball, 29-30, 144
Bathing
    skipping baths, 6
    water footprint of, 120, 136
Bathrooms
    airport restrooms, 36-37, 147
    baths, skipping, 6
    brushing teeth, 5, 120, 135
    high-efficiency faucets, 24
    low-flush or dual flush toilet, 5,
        24, 142-43
    references, 135-36, 142-43
    running faucets, 23-24, 142
    shaving, 6, 120
    showerhead, low-flow, 6, 136
    showers
        at hotels, 37
        length of, 6, 136, 145
        skipping extra, 28, 30-31
    at sports venues, 29-30
    waterless urinals, 24, 143
    water-saving ideas for, 5-6
    workplace, 23-24, 142-43
Beans, 60, 159
Beds
    mattress, 75
    pets, 99, 179
    pillows, 76, 169
    references, 166-67
Beef, 54-55, 155

Beer, 64, 119, 160-61
Beets, sugar, 54
Bell pepper, 53
Berries
    blueberries, 49-50, 152
    grapes, 50, 152
    raisins, 50, 152
    raspberries, 50, 153
    references, 152-53
    strawberries, 50-51, 153
Beverages
    alcoholic
        beer, 64, 160-61
        gin, 64, 161
        rum, 64, 161-62
        tequila, 64-65, 162
        vodka, 65, 162
        whiskey, 65, 162
        wine, 64, 65, 119, 161, 162, 177
    aluminum cans, 61, 159
    bottled water, 119
    coffee
        fair-trade, 62
        references, 136, 142, 160
        water footprint calculation,
            114-15
        water use for, 6, 22-23, 43, 62,
            128-29
    glass bottles, 61, 159
    juices
        apple, 62, 160
        orange, 62, 160
        pineapple, 63, 160
        tomato, 63, 160
        water footprint calculation,
            114-15
    references, 159-60
    soda, 62, 159
    tea, 62, 114-15, 160
Black pepper, 48
Black tea, 62, 114-15, 160
Blankets, 75, 167
Blueberries, 49-50, 152
Boats, 91-92, 94, 95, 178
Books, 87, 173-74

Bottled water, 119
Bottles, glass, 61, 159
Bowls, pets, 99-100, 179
Bran, wheat, 59, 158
Breakfast, 114-15
Break room, 23, 142
Broccoli, 51, 153
Bromelain, 48-49
Brushing teeth, 5, 120, 135
Bubbler, 23, 142
Building materials and appliances
    cabinetry, 105, 181-82
    carpeting, 105, 182
    clothes dryers, 105-6, 182-83
    granite countertops, 106, 183
    gutters, 106, 183-84
    hammer, 181
    hardwood flooring, 106, 184
    insulation, 106-7, 184
    linoleum flooring, 107, 109, 184
    ovens and stoves, 107, 184
    piping, 107, 184-85
    references, 181-87
    refrigerators, 107, 185
    roofing, 107-8, 185
    stone-tile flooring, 108, 185
    tile countertops, 108, 186
    tool belt, 181
    washing machines, 108, 186
    water footprint of, 103-9
    windows, 109, 186-87
Businesses, water-saving ideas for,
    21-26
Butter, 56, 156

C

Cabbage, 51, 153
Cabinetry, 105, 181-82
Candy, 63, 160
Cans, aluminum, 61, 159
Cantaloupe, 47
Carpeting, 105, 182
Carrots, 51, 153
Cars, 92-93, 123, 177
Cat litter, 98

Celery, 51–52, 153
Cereals
    cornflakes, 58–59, 158
    flax, 59, 158
    granola, 59, 158
    references, 158
    water footprint calculation, 114
    water footprint of, 58–59
    wheat bran, 59, 158
Chairs, 74, 75, 167
Cheese, 56, 156
Chest, 76, 168
Chicken, 55, 117, 155
Cholesterol, 56
Clean Water Act, 125–26
Clothes dryers, 105–6, 182–83
Clothing
    jackets, 69, 163
    jeans and pants, 67, 68, 69,
        163–64
    references, 163–66
    running sneakers, 69–70, 164
    second-hand, 72
    shirts, 70, 164–65
    shoes, 70, 165
    socks, 70, 165
    suits, 70, 165
    sweaters, 71, 166
    T-shirts, 71, 166
    underwear, 71, 166
    water footprint of, 67–72, 122–23
Coconuts, 49, 152
Coffee
    fair-trade, 62
    references, 136, 142, 160
    water footprint calculation,
        114–15
    water use for, 6, 22–23, 43, 62,
        128–29
Collars, pet, 100, 179
Colorado River, 125
Composting, 7, 18
Computers, 87, 174
Cooking, water use in, 6–7, 136
Cop, water, 127

Copper
    gutters, 183
    pipes, 107, 184–85
Copying, 85
Corn, 51–52, 153
Cornflakes, 58–59
Cosmetics, 81, 171
Cotton, 68, 70–72
Couches, 76, 168
Countertops
    granite, 106, 183
    tile, 108, 186
Court tiles, 31, 145
Cucumbers, 52, 154

D
Dairy products
    butter, 56, 156
    cheeses, 56, 156
    eggs, 56, 156
    milk, 56–57, 156
    references, 156
    yogurt, 57, 156
Dehydration, 30, 39
Diamonds, 91
Dinner, water footprint of, 116–18
Dinnerware, 76, 168
Dishwasher, 8, 137
Dishwashing
    rinsing dishes, 8
    water footprint of, 121
Disposal, garbage, 7, 137
Drains, maintaining, 20
Drinking fountain, 32, 36
Drinking water, 7, 137
Drinks. See Beverages
Dry farming, 64

E
Easter Island, 128
E-books, 88
Ecosystem, connectivity in, 126
Eggs, 56, 114, 156
Elevation, dehydration and, 39
E-mail, 88

Environmental Protection Agency
    (EPA), 67, 68
Erasers, 87, 174
Erosion, 39
Eureka lemon, 47
Evaporative cooler, 9
Exercise
    morning workouts, 30-31, 32
    sports, 27-32, 144-46
Exteriors, 139

F

Faucets
    aerators, 9, 11, 138
    leaks, 9, 137
    left running in restrooms, 23-24,
        142
    sensors for, 24, 142
    WaterSense labeled, 24, 142
Feces, pet, 98, 101
Fences, as windbreaks, 19
Fertilizer, choice of, 18
Flax, 59
Flooring
    hardwood, 106, 184
    linoleum, 107, 109, 184
    stone-tile, 108, 185
Flour, white, 58, 157-58
Food and beverage products
    beans, 60-61, 159
    berries, 49-51, 152-53
    beverages, 61-62, 159-60
    buying in season, 66
    buying local, 66
    cereals, 58-59, 158
    dairy products, 56-57, 156
    fruits, 45-49, 149-51
    general recommendations for
        choosing, 65-66
    grains, 57-58, 156-58
    juices, 62-63, 160
    labels, 45, 149
    meats, 54-56, 155-56
    nuts, 49, 151-52
    for pets, 100, 101, 180

pizza and pasta, 60, 159
references, 148-62
snacks, 63-64, 160
vegetables, 51-54, 153-55
wastage, 25, 26, 45, 143
water footprint of, 43-66, 114-19
wine and spirits, 64-65, 160-62
Footprint, calculating. See Water
    footprint calculator
Forest Stewardship Council (FSC),
    74, 105, 106, 109
Fountains
    drinking, 32, 36
    garden, 17, 140
Freshwater
    agricultural use, 43-45, 148
    amount available on Earth, 126
Fruit
    apples, 45
    apricots, 45-46, 149
    avocados, 46, 149
    bananas, 46, 149-50
    buying in season, 66
    grapefruit, 46, 150
    kiwifruit, 46, 150
    lemons, 47, 150
    limes, 47, 150
    mangoes, 47, 150
    melons, 47-48, 150
    nectarines, 48
    oranges, 48
    peaches, 48, 150-51
    pears, 48, 151
    pepper, black, 48, 151
    pineapples, 48-49, 151
    plums, 49, 151
    references, 149-51
    water footprint calculations, 114,
        118
FSC, 74, 105, 106, 109
Furnishings
    beds, 75, 166-67
    blankets, 75, 167
    chairs, 75, 167
    chest, 76, 168

couches, 76, 168
dinnerware, 76, 168
glassware, 76, 168-69
pillows, 76, 169
pots and pans, 77, 78, 169
references, 166-70
rugs, 77, 170
second-hand, 74-75
sheets, 77, 170
silverware, 77-78, 170
tables, 78, 170
water footprint of, 73-78

G
Garbage disposal, 7, 137
Garden hoses, 16, 140
Gardens
fertilizer choice, 18
soil drainage, 17-18
water footprint of, 121
watering, 17, 18
Garlic, 52, 154
Gin, 64, 161
Glass bottles, 61, 159
Glassware
washing, 7-8
water footprint of, 76, 168-69
Golf courses, 28, 29, 144
Grains
barley, 57, 156
oats, 57, 157
references, 156-58
rice, 58, 157
rye, 58, 157
wheat, 58, 157
white flour, 58, 157-58
Granite countertops, 106, 183
Granola, 59
Grapefruit, 46, 150
Grapes, 50, 152
Grass, cutting height, 16, 139
Gray water recapture, 15-16, 139
Great Lakes, 125
Great Plains (Ogallala) aquifer, 125
Green tea, 62

Green water, 74
Gutters, 15, 106, 139, 183-84
Gymnasiums, 32, 145-46

H
Hammer, 181
Hardwood flooring, 106, 184
Health and beauty products
antiperspirants, 80-81, 171
cosmetics, 81, 171
moisturizers, 81, 171-72
packaging, 79-80, 83
perfumes, 81-82, 172
pollution from, 80, 83
references, 170-73
shampoos, 82, 172
soaps, 82, 172
spending on, 79
toothpaste, 82-83, 173
water-saving ideas for, 79-83
Hiking, 38-39, 148
Home. See also Building materials
and appliances; Furnishings
bathroom, 5-6, 135-36
daily water use amount, 3
indoor plumbing, 4
kitchen, 6-8, 136-37
laundry room, 8-9, 137
living areas, 9-10, 137-38
references, 135-38
size of average US home, 3-4
water footprint of, 3-12, 120-22
Honey, 54
Hose, garden, 16, 140
Hotels, 33, 34, 37-38, 40, 147-48
Hot tub, 38, 148
Hydration, 30, 145
Hyponatremia, 30

I
Ice, 7, 137
Iceberg lettuce, 52
Ice-skating, 31, 145
Indian Woman Yellow beans, 60
Industry, water use by, 21-22

Information resources
    general information and news,
        131
    to take action, 132
    virtual water, 131
    water conservation solutions,
        131-32
    water policy and management,
        132
    water research, science, and
        technology, 132
Insulation
    fiberglass, 106, 184
    pipe, 10, 138
    shredded newspaper, 106-7, 184
Irrigation, 13, 25-26, 35, 143, 146

J
Jackets, 69, 163
Jacob's Cattle beans, 60
Jeans, 67, 68, 69, 163-64
Jets, 92, 93, 177
Jewelry, 93, 95, 177
Juices
    apple, 62
    orange, 62
    pineapple, 63
    references, 160
    tomato, 63
    water footprint calculation,
        114-15

K
Kitchens
    appliances
        dishwasher, 8, 137
        ovens and stoves, 107, 184
        refrigerators, 107, 185
    coffee making, 6, 136
    composting, 7
    cooking, water use in, 6-7, 136
    dinnerware, 76, 168
    drinking water, 7, 137
    garbage disposal, 7, 137
    glassware, 7-8, 76, 168-69

ice cubes, 7, 137
    pots and pans, 77, 78, 169
    references, 136-37
    rinsing dishes, 8
    vegetable preparation, 7
    washing glassware, 7-8
    water-saving ideas for, 6-8
Kiwifruit, 46, 150

L
Lamb, 55, 155
Landscaping, 143
Laundry
    hotel laundry service, 34, 38, 147
    water footprint of, 121
Laundry room, 8-9, 137
Lawn and garden areas, 139-40
    fertilizer choice, 18
    golf courses, 28, 29, 144
    grass, cutting height, 16, 139
    irrigation, 13, 25-26, 143
    office landscape, 25-26, 143
    overwatering, 19-20
    soil drainage, 17-18, 140
    sprinklers
        office landscape use, 26
        proper use of, 16
        timers, 16, 34, 35, 140
    time of day for watering, 16
    water footprint of, 121
    watering gardens, 17
    xeriscaping, 17, 20, 121, 140
Laws, water, 127-28
Leaks
    amount of water lost by, 20, 34
    checking for, 9
    faucets, 9, 137
    toilet, 9, 138
    while traveling, 34
Leashes, 100, 180
Leather, 69, 70, 72, 75-76
Lemons, 47, 150
Lettuce, 52, 154
Licorice, 63, 128
Lifestyle, water footprint of, 122-23

Limes, 47, 150
Linoleum flooring, 107, 109, 184
Living areas
    faucet aerators, low-flow, 9
    leaks, checking for, 9
    pipe insulation, 10
    plant watering, 9
    references, 137-38
    tankless water heaters, 10
    water meters, 10
    water-saving ideas for, 9-10
Logging, 74
Lunch, water footprint of, 115-16
Luxury items
    cars, 92-93, 123, 177
    jets, 92, 93, 177
    jewelry, 93, 95, 177
    references, 176-78
    televisions, 93-94, 178
    watches, 94, 178
    water footprint of, 91-95
    wine, 64, 65, 119, 177
    yachts, 91-92, 94, 95, 178

M
Mangoes, 47, 150
Mattress, 75
Meals, water with, 37
Meats
    beef, 54-55, 155
    chicken, 54-55, 155
    lamb, 55, 155
    pork, 55, 155
    references, 155-56
    turkey, 56, 156
    water footprint of, 54-56
Medium-density fiberboard (MDF), 105, 182
Meetings
    references, 143
    water-saving ideas for, 24-25
Melons, 47-48, 150
Meyer lemon, 47
Milk, 56-57, 156
Moisturizers, 81, 171-72

Mushrooms, 52-53, 154
Muskmelon, 47

N
Nectarines, 48
Newspaper, as insulation, 106-7, 184
Nuts
    almonds, 49, 151-52
    coconuts, 49, 152
    references, 151-52

O
Oatmeal, 57
Oats, 57, 157
Office products. See School and office products
Ogallala (Great Plains) aquifer, 125
Omega-3 fatty acids, 59
Onions, 53, 128, 154
Orange juice, 62, 160
Oranges, 48
Outdoors
    exteriors, 15-16, 139
    lawn and garden areas, 16-18, 139-40
    pools, 18-19, 140-41
    references, 138-41
    water-saving ideas for, 15-20
Ovens and stoves, 107, 184
Overhydration, 30

P
Pans. See Pots and pans
Pants, 69, 163-64
Paper
    copying, 85
    recycling, 87-88, 89
    references, 142, 174
    water footprint calculation, 123
    water use in production of, 86-88
    workplace use, 22
Pasta, 6-7, 60, 116, 128, 159
Peaches, 48, 150-51

Pears, 48, 151
Peas, 53, 154
Pencils, 86, 88, 174-75
Pens, 88, 175
Pepper, black, 48, 151
Peppers, bell, 53, 154
Perfumes, 81-82, 172
PET, 61
Pets
   beds, 99, 179
   bowls, 99-100, 179
   collars, 100, 179
   food, 100, 101, 180
   leashes, 100, 180
   number in United States, 97
   poop bags, 101, 180-81
   references, 178-81
   toys, 100, 101, 180
   treats, 101, 180
   water footprint of, 97-101
Pillows, 76, 169
Pineapple juice, 63, 160
Pineapples, 48-49, 151
Pipes
   burst, 34, 35, 146
   copper, 107, 184-85
   insulation, 10, 138
Pizza, 59, 159
Plants. See also Lawn and garden
      areas
   care of while away from home,
      35-36
   drought-resistant, 17
   overwatering, 19-20
   watering indoor, 9, 35-36, 121,
      137, 146
Plastic, water footprint of, 123
Plastic water bottles
   at airports, 36, 146-47
   amount of water required to
      produce, 32, 61, 143, 159
   at gyms, 32
   at hotels, 37
   at travel destination, 38
   using alternatives to, 24-25, 30

Plums, 49, 151
Pollution
   Clean Water Act, 125-26
   cosmetics disposal, 80, 83
   pet waste, 98
Polyethylene terephthalate (PET), 61
Pools
   covers, 19, 141
   draining, 19, 141
   filters, 19
   references, 140-41
   splashing, avoiding, 18-19, 29
   water footprint calculation, 122
   water neutrality, 18
   water-saving ideas for, 18-19
   windbreaks, 19, 141
Poop bags, 180-81
Popcorn, 63, 160
Pork, 55, 117, 155
Potato chips, 63-64, 118-19, 160
Potatoes, 53, 154-55
Pots and pans, 77, 78, 169
Printers, 88, 175-76
Probiotic foods, 57

Q
Quick guide, to virtual water
      content, 189-96

R
Rain barrels, 14, 15, 139
Rainwater, capture of, 14, 15, 139
Raisins, 50, 152
Raspberries, 50, 153
Recycling
   aluminum cans, 61
   paper, 89
Red meat, 54-55, 117
Red wine, 64, 161
References
   building materials and
      appliances-related, 181-87
   cabinetry, 181-82
   carpeting, 182
   clothes dryers, 182-83

granite countertops, 183
gutters, 183-84
hammer, 181
hardwood flooring, 184
insulation, 184
linoleum flooring, 184
ovens and stoves, 184
piping, 184-85
refrigerators, 185
roofing, 185
stone-tile flooring, 185
tile countertops, 186
tool belt, 181
washing machines, 186
windows, 186-87
clothing-related, 163-66
    jackets, 163
    jeans and pants, 163-64
    running sneakers, 164
    shirts, 164-65
    shoes, 165
    socks, 165
    suits, 165
    sweaters, 166
    underwear, 166
food and beverage-related,
    148-62
    beans, 159
    berries, 152-53
    beverages, 159-60
    cereal, 158
    dairy, 156
    fruits, 149-51
    grains, 156-58
    juices, 160
    meats, 155-56
    nuts, 151-52
    pizza and pasta, 159
    snacks, 160
    vegetables, 153-55
    wine and spirits, 160-62
furnishings-related, 166-70
    beds, 166-67
    blankets, 167
    chairs, 167

chest, 168
couches, 168
dinnerware, 168
glassware, 168-69
pillows, 169
pots and pans, 169
rugs, 170
sheets, 170
silverware, 170
tables, 170
health and beauty-related,
    170-73
    antiperspirants, 171
    cosmetics, 171
    moisturizers, 171-72
    perfumes, 172
    shampoos, 172
    soaps, 172
    toothpaste, 173
home-related, 135-38
    bathroom, 135-36
    kitchen, 136-37
    laundry room, 137
    living areas, 137-38
luxury-related, 176-78
    cars, 177
    jets, 177
    jewelry, 177
    televisions, 178
    watches, 178
    wine, 177
    yachts, 178
outdoors-related, 138-41
    exteriors, 139
    lawn and garden areas, 139-40
    pools, 140-41
pet-related, 178-81
    beds, 179
    bowls, 179
    collars, 179
    food, 180
    leashes, 180
    poop bags, 180-81
    toys, 180
    treats, 180

References *(cont.)*
    school and office products-
        related, 173-76
        books, 173-74
        computers, 174
        erasers, 174
        paper, 174
        pencils, 174-75
        pens, 175
        printers, 175-76
        tape, 176
    sports-related, 144-46
        summer sports, 144-45
        winter sports, 145-46
    travel-related, 146-48
        airport, 146-47
        destination, 148
        hotel, 147-48
        leaving home, 146
    work-related, 141-43
        break room, 142
        landscaping, 143
        meetings, 143
        paper, 142
        restroom, 142-43
Refrigerators, 107, 185
Restrooms. *See* Bathrooms
Rice, 58, 116, 157
Riparian law, 14
Romaine lettuce, 52
Roof, as catchment area for water,
    15
Roofing, 107-8, 139, 185
Rugs, 77, 170
Rum, 64, 161-62
Running, 30, 144
Running sneakers, 69-70,
    164
Rye, 58, 157

S

Salad, 115-16
Salami, 55
Sandwich, 115
Sausage, 55

School and office products
    books, 87, 173-74
    computers, 87, 174
    erasers, 87, 174
    paper, 85-89, 174
    pencils, 86, 88, 174-75
    pens, 88, 175
    printers, 88, 175-76
    references, 173-76
    tape, 88, 176
    water footprint of, 85-89
Shampoos, 82, 172
Shaving, 6, 120, 136
Sheets, 77, 170
Shirts, 70, 164-65
Shoes, 70, 165
Shower
    at hotels, 37
    length of, 6, 136, 145
    skipping extra, 28, 30-31
Showerhead, low-flow, 6, 136
Shutoff switch, main water valve,
    35, 40
Silverware, 77-78, 170
Ski resorts, 31, 145
Snacks
    candy, 63, 160
    popcorn, 63, 160
    potato chips, 63-64, 160
    references, 160
    water footprint of, 118-19
Sneakers, running, 69-70, 164
Soaps, 82, 172
Socks, 70, 165
Soda, 62, 116, 159
Soil, types and drainage, 17-18, 140
Soybeans, 61, 159
Spa, 38, 148
Spinach, 53, 155
Sports
    professional sports, 27-28
    references, 144-46
    summer sports, 28-31, 144-45
    water-saving ideas for, 27-32
    winter sports, 31-32, 145-46